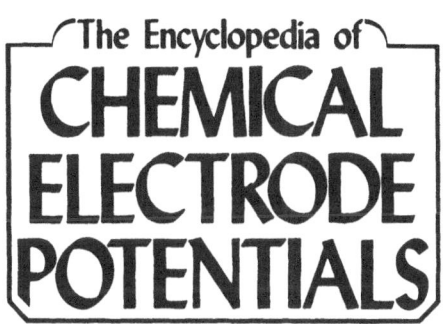

The Encyclopedia of
CHEMICAL
ELECTRODE
POTENTIALS

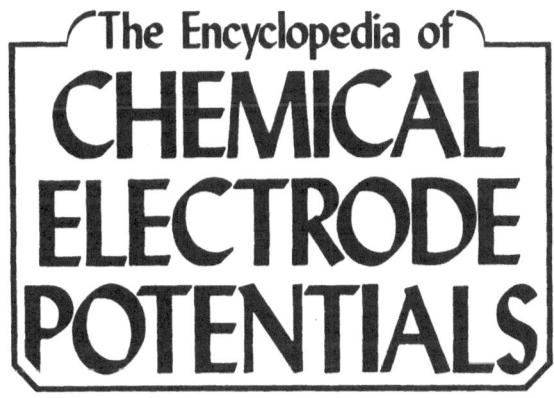

The Encyclopedia of CHEMICAL ELECTRODE POTENTIALS

Marvin S. Antelman

Chemical Consultant and President
Antelman Technologies
Chairman of the Board, Tilaco Chemicals Ltd.
Providence, Rhode Island

With the assistance of

Franklin J. Harris, Jr.

Director of Research, Tilaco Chemicals Ltd.
Providence, Rhode Island

PLENUM PRESS - NEW YORK AND LONDON

Library of Congress Cataloging in Publication Data

Antelman, Marvin S., 1933-
  The encyclopedia of chemical electrode potentials.

  Bibliography: p.
  Includes index.
  1. Electrodes—Tables. 2. Electrochemistry—Tables. I. Harris, Franklin J. II. Title.
QD571.A64   541.3'724   81-20986
ISBN-13: 978-1-4613-3376-0   e-ISBN-13: 978-1-4613-3374-6   AACR2
DOI: 10.1007/978-1-4613-3374-6

© 1982 Plenum Press, New York

Softcover reprint of the hardcover 1st edition 1982

A Division of Plenum Publishing Corporation
233 Spring Street, New York, N.Y. 10013

This book is dedicated to the memory of
Professor Bruno Z. Kisch
(1890-1966)

Bruno Z. Kisch was a noted medical authority and Professor of Chemistry at Yeshiva University, my alma mater. I had the privilege of studying under Professor Kisch, particularly his Philosophy of Science course, in which he demonstrated the erudition of a true Renaissance man. He was an authority in the field of philosophy and exerted a profound influence on Jewish theology, especially in areas where science, religion, and medical ethics interact.

Dr. Kisch was an outstanding chemist and utilized his comprehensive knowledge of chemistry to make many advances in the field of cardiology. The *Journal of the American Medical Association* recognized his creative contributions by dedicating an issue to him on the occasion of his 70th birthday. Kisch also made advances in the fields of electron microscopy and biochemistry, and was a founder and president of the American College of Cardiology.

One of his more famous medical treatises was *Microscopy of the Cardiovascular System* (1960). In the 1930's he also published several books dealing with religion and science, e.g., *Gottesglaube und Naturerkenntnis* (1936), in his native Germany. He escaped the Holocaust to the U.S.

Despite his active schedule as a teacher, author, and researcher, Professor Kisch maintained a medical practice in midtown Manhattan. I was a privileged visitor to the private science and medical library which he maintained there and consulted him frequently about various ideas and concepts in chemical research. I shall always be grateful to Professor Kisch for his support and guidance.

Marvin S. Antelman

ACKNOWLEDGMENT

I wish to express my thanks to the people who were instrumental in making the production of this encyclopedia a reality.

First, to the late Mark Weisberg (1889-1963) of Providence, Rhode Island, who was the founder of Alrose Chemical Co., which was later acquired by Geigy. He was a pioneer in the field of chelating agents. In 1958, after having sold Alrose to Geigy, he, together with Bradley Dewey, formed the Hampshire Chemical Co. (a division of Grace). At the same time, Mark Weisberg also formed a firm devoted to metal finishing and precious metal plating products by the name of Technic, which today is situated in Cranston, Rhode Island, with his son Alfred Weisberg as its President. In 1958, I was privileged to join Mark Weisberg as Chief Applications Chemist for Hampshire. He first made me aware of the need for EMF data to develop strippers and inhibitors for electroplated finishes and for the utilization of the chemical chelating agents that Hampshire Chemical was manufacturing.

I also wish to thank Franklin J. Harris, Jr., Director of Research of Tilaco Chemicals Ltd., for his laborious and painstaking efforts in aiding me in gathering information and calculating many of the EMF values included in this encyclopedia.

Acknowledgment is also due to Beverly Fairhurst for her work in preparing the manuscript for publication and for her careful proofreading. Beverly Fairhurst has had experience in science education and was on the research staff of Tilaco for many years.

Credit is also due to Ann Fine for aiding in the compilation of this work and performing some of the proofreading and stenographic tasks involved in making this work a reality.

Last but not least, I would like to thank my dear wife, Sylvia Joyce, without whose support and forbearance this work would have been impossible.

January 1982   Marvin S. Antelman

CONTENTS

INTRODUCTION

This book is an attempt to compile the most comprehensive listing of chemical electrode potentials to date.

In compiling this encyclopedia, many papers, articles, and publications—often with contradictory data—were consulted. Of all the published material available, the authors found the following of particular interest:

- G. Milazzo and S. Caroli, *Tables of Standard Electrode Potentials,* John Wiley and Sons, Inc. (1978).
- A. J. deBethune and N. A. Swendeman Loud, *Standard Aqueous Electrode Potentials and Temperature Coefficients at 25°C,* Clifford A. Hempel (1964).
- W. Mansfield Clark, *Oxidation-Reduction Potentials of Organic Systems,* The Williams and Wilkins Co. (1960).

However, it was our intent also to include unpublished complex formation EMF data. Furthermore, when we consulted published EMF data, we found that the compilations often paid no attention to disparate conditions and varying pressures and electrolyte concentrations.

To achieve a useful compilation, we resorted to Nernst equation thermodynamic calculations to reconcile disparate data. We also utilized the following relationships to calculate EMF values for inclusion in our tables from association constants:

$$\ln K = \frac{nFE}{RT} \qquad \log K = -nE(16.9) \qquad E_2 = E^0 + E_1$$

As a practical example, consider the reaction

$$Ca + Nta^{-3} - 2e = Ca(Nta)^-$$

We may write

$Ca - 2e = Ca^{+2}$	$\underline{E^0 = -2.87}$
$\underline{Ca^{+2} + Nta^{-3} = Ca(Nta)^-}$	$E_1 = ?$
$Ca - 2e + Nta^{-3} = Ca(Nta)^-$	$E_2 = ?$

Given log K Ca(Nta)$^-$ = 6.41,

$$6.41 = -(2)(16.9)E_1 \qquad \underline{E_1 = -0.19}$$
$$E_2 = E^0 + E_1$$
$$= -2.87 + (-0.19) = -3.06$$

The two most useful data compilations with bibliographic references on complexes and acids and bases were:

Adrien Albert and E. P. Serjeant, *Ionization Constants of Acids and Bases,* John Wiley and Sons, Inc. (1962).

K. B. Yatsimirskii and V. P. Vasil'ev, *Instability Constants of Complex Compounds,* Consultants Bureau (1960).

We have utilized their data to calculate EMF values that could be integrated into our tables.

There is one point of nomenclature that should be emphasized. Many electrochemists differentiate strictly between the terms *potential* and *electromotive force.* Generally potential refers to an experimentally measured quantity whereas electromotive force is a thermodynamic quantity obtained by calculation or derived from thermodynamic relationships. We have intermingled the two in constructing our electrochemical series, referring to both simply as EMF data.

In preparing this work and in relying on both calculated and experimental data, we have had the applied chemist in mind more than the pure chemist. While the latter can certainly utilize our data, we have found that most industrial chemists refer to chemical potential data for guidance in the day-to-day performance of their experiments. Real laboratory conditions and the experimental variables involved, especially the presence of other macrochemicals or microimpurities, will tend to drastically alter actual electrode potentials of various systems.

For maximum utility the encyclopedia has been arranged into five sections. Part I is an electromotive series. Part II lists tables of EMF values according to elements. Part III lists EMF values of common inorganic anions. Part IV lists electrode potentials of various organic functional groups. Finally, Part V lists EMF data for the inorganic nitrogen bases, NH_3 and N_2H_4.

To make the book easier to use by fitting our data neatly into tabular form, we had to introduce many abbreviations, which are explained at the end of the volume.

HOW TO USE THE TABLES

The tables are arranged according to reduction potentials. While the usual convention is to speak of the oxidized state accepting electrons to yield the reduced state,

$$Ox. + e = Red.$$

or, in the case of zinc,

$$Zn^{+2} + 2e = Zn$$

we have chosen to think of the reduced state as losing electrons and changing to the oxidized state. The sign convention is that of the reduction potential; for example,

$$Red. - e = Ox.$$

or, in the case of zinc,

$$Zn - 2e = Zn^{+2} \qquad Reduction\ Potential = -0.76.$$

If we were to use oxidation potentials, the value would be $+0.76$.

The symbol for the electron can be either e or e⁻; we have used the e throughout. The H^+ notation is used in preference to hydronium or positronium.

Part I. Electrochemical Series

The electrochemical series (for the first time) employs a systematic subclassification arrangement which differentiates anions, cations, complexes, and compounds. The range of the series is from the most negative reduction potential of -4.88 for the entity BH, through the reduction potentials below hydrogen, to that of $+3.06$ for the reduction of hydrofluoric acid to elemental fluorine.

The columns following the EMF value are Element, Anion, Compound or Anion, Cation, and Water, on the reduction side. This arrangement is of particular interest for it will enable the user to classify reactions quickly. The reduction side and the oxidation side are separated by an "equals" sign, with the oxidized state to the right of the sign. We have endeavored to keep most of the simple cations in the Cation column on the oxidation side of the series, while most complex ions are classified with the compounds. Compounds also include alloys, such as zinc-mercury amalgam.

While some electromotive series differentiate between reactions in the acid and basic ranges, our tables are all-encompassing, as the basic reactions usually involve OH^-. The OH^- reactions are considered to be at a pH of 11, other reactions in the absence of the H ion are considered to have a pH of 7, and those involving the H ion a pH of 1. Those involving both H^+ and OH^- can be construed to occur at a pH of 7.

The electrochemical series gives half-reactions, and it is a simple matter to combine the half-reactions of two different potentials to obtain the full reaction.

Part II. Chemical Electrode Potentials by Element

This section utilizes the same convention as the electrochemical series of Part I, with the same headings, except that the arrangement of the potentials is by element in alphabetical order according to symbol.

Part III. Chemical Electrode Potentials of Inorganic Anions

In this section, the arrangement is identical to Parts I and II, except that the Anion column follows the EMF value and the arrangement is in alphabetical order of the full names of inorganic anions.

Part IV. Chemical Electrode Potentials of Organic Functional Groups

This part contains different functional groups, with an Element column, usually hydrogen, followed by a Compound column. In the oxidized state, the columns are Compound and Cation. The cation is usually H^+ if present.

Part V. Chemical Electrode Potentials of Inorganic Nitrogen Bases

This section is self-explanatory.

Abbreviations

This section lists commonly used abbreviations for various chemical entities, e.g., Edta, which is the common abbreviation for ethylenediaminetetraacetic acid. Some of these abbreviations are used for acids and bases as well as their complexes. Thus, Edta is used for the acid as well as its complex, Lac for lactates and lactic acid, and Gu for guanidine complexes as well as guanidine. Where space allows, short names such as lactate are written out in full. To meet space limitations some special abbreviations had to be introduced, e.g., DCPIC for 2,6-dichlorophenol-indo-o-cresol.

Abbreviations commonly used for state designations are not included in the list. The reader should have no difficulty in recognizing abbreviations such as (g) for gaseous and (aq) for aqueous. Alpha, beta, gamma, and delta are abbreviated as (a), (b), (gm), and (D).

PART I

ELECTROCHEMICAL SERIES

EMF	Elem	Anion	Compound or anion	Cation	H_2O	e =	Cation	Anion	Complex or compound	Compound or element
-4.88			BH			-e	H^+			B
-4.10	Sr					-e	Sr^+			
-4.00	Ra					-e	Ra^+			
-3.80	Ca					-e	Ca^+			
-3.40	Eu					-2e	Eu^{+2}			
-3.40			$HN_3(g)$			-e	H^+			$3/2N_2$
-3.34	Li		$Amac^{-3}$			-e			$LiAmac^{-2}$	
-3.23	Ca		$Data^{-4}$			-2e			$CaData^{-2}$	
-3.21	Li		Nta^{-3}			-e			$LiNta^{-2}$	
-3.19	Li		$Edta^{-4}$			-e			$LiEdta^{-3}$	
-3.18	Ca		$Edta^{-4}$			-2e			$CaEdta^{-2}$	
-3.15	Li		Saa^{-3}			-e			$LiSaa^{-2}$	
-3.15	Sr		$Edta^{-4}$			-2e			$SrEdta^{-2}$	
-3.13	Ba		$Edta^{-4}$			-2e			$BaEdta^{-2}$	
-3.13	Ca		$Amac^{-3}$			-2e			$CaAmac^-$	
-3.12	Sm					-2e	Sm^{+2}			
-3.12	Sr		$Amac^{-3}$			-2e			$SrAmac^-$	
-3.11	Ca		Hed^{-3}			-2e			$CaHed^-$	
-3.11	Ca		Nta^{-3}			-2e			$CaNta^-$	
-3.10	Ba		$Amac^{-3}$			-2e			$BaAmac^-$	
-3.09	Ba		Nta^{-3}			-2e			$BaNta^-$	
-3.09			$HN_3(aq)$			-e	H^+			$3/2N_2$
-3.09	Sr		Nta^{-3}			-2e			$SrNta^-$	
-3.08	Ca		$Tmta^{-4}$			-2e			$CaTmta^{-2}$	
-3.07	K		$(FeCy)^{-4}$			-e			$K(FeCy)^{-3}$	
-3.05	Ba		$P_4O_{12}^{-4}$			-2e			$BaP_4O_{12}^{-2}$	
-3.05	Li					-e	Li^+			
-3.05	Li		$Lactate^{-2}$			-e			$Li(Lactate)^-$	
-3.04	Ca		Esb^{-3}			-2e			$CaEsb^-$	
-3.04	Sr		$P_4O_{12}^{-4}$			-2e			$SrP_4O_{12}^{-2}$	
-3.04	Sr		$Tmta^{-4}$			-2e			$SrTmta^{-2}$	
-3.03	Ba		$Tmta^{-4}$			-2e			$BaTmta^{-2}$	
-3.03	Ba		$(FeCy)^{-4}$			-2e			$Ba(FeCy)^{-2}$	

EMF	Elem	Anion	Compound or anion	Cation	H_2O	e =	Cation	Anion	Complex or compound	Compound or element
-3.03	Ca		Ers^{-3}			-2e			$CaErs^-$	
-3.03	Ca		Esa^{-3}			-2e			$CaEsa^-$	
-3.03	Ca		Est^{-3}			-2e			$CaEst^-$	
-3.03	Ca		$Ist(b)^-$			-2e			$Ca(Ist(b))^+$	
-3.03	Ca		$Met(b)^-$			-2e			$Ca(Met(b))^+$	
-3.03	Ca		$P_4O_{12}^{-4}$			-2e			$CaP_4O_{12}^{-2}$	
-3.03	Li		OH^-			-e			$LiOH$	
-3.03	Li		Gly^{-2}			-e			$Li(Gly)^-$	
-3.02	Ca		$2OH^-$			-2e			$Ca(OH)_2$	
-3.02	Ca		$2Amac^{-3}$			-2e			$Ca(Amac)_2^{-4}$	
-3.02	Ca		$Ist(a)^-$			-2e			$Ca(Ist(a))^+$	
-3.02	Ca		$Met(a)^-$			-2e			$Ca(Met(a))^+$	
-3.02	Ca		$P_2O_7^{-4}$			-2e			$CaP_2O_7^{-2}$	
-3.01	Ca		$Himda^{-2}$			-2e			$CaHimda$	
-3.01	Ca		Saa^{-3}			-2e			$CaSaa^-$	
-3.01	K		$(CoCy)^{-3}$			-e			$K(CoCy)^{-2}$	
-3.01	Ca		$Trop^-$			-2e			$CaTrop^+$	
-3.01	K		$(FeCy)^{-3}$			-e			$K(FeCy)^{-2}$	
-3.00	Ba		$P_3O_9^{-3}$			-2e			$BaP_3O_9^-$	
-3.00	Ca		Coy^-			-2e			$CaCoy^+$	
-2.99	Ba		$2OH^-$		$8H_2O$	-2e			$Ba(OH)_2 \cdot 8H_2O$	
-2.99	Ba		Cit^{-3}			-2e			$BaCit^-$	
-2.99	Sr		$P_3O_9^{-3}$			-2e			$SrP_3O_9^-$	
-2.99	Sr		Saa^{-3}			-2e			$SrSaa^-$	
-2.98	Ca		$2Ist(a)^-$			-2e			$Ca(Ist(a))_2$	
-2.98	Ca		$2Ist(b)$			-2e			$Ca(Ist(b))_2$	
-2.98	Ca		$2Met(b)^-$			-2e			$Ca(Met(b))_2$	
-2.98	K		SO_4^{-2}			-e			KSO_4^-	
-2.98	Ra		Cit^{-3}			-2e			$RaCit^-$	
-2.98	Sr		Cit^{-3}			-2e			$SrCit^-$	
-2.97	Ba		Ox^{-2}			-2e			$BaOx$	
-2.97	Ba		$S_2O_3^{-2}$			-2e			BaS_2O_3	

EMF	Elem	Anion	Compound or anion	Cation	H_2O	e =	Cation	Anion	Complex or compound	Compound or element
-2.97	Ba		Saa^{-3}			-2e			$BaSaa^-$	
-2.97	Ca		$2Met(a)^-$			-2e			$Ca(Met(a))_2$	
-2.97	Ca		$2Nta^{-3}$			-2e			$Ca(Nta)_2^{-4}$	
-2.97	Ca		$P_3O_9^{-3}$			-2e			$CaP_3O_9^-$	
-2.97	K		$S_2O_3^{-2}$			-e			$KS_2O_3^-$	
-2.97	K	ReO_4^-				-e				$KReO_4$
-2.97	Sr		Ox^{-2}			-2e			$SrOx$	
-2.97	Sr	$(FeCy)^{-3}$				-2e			$Sr(FeCy)^-$	
-2.96	Ca		Cit^{-3}			-2e			$CaCit^-$	
-2.96	Ca		$HCit^{-2}$			-2e			$CaHCit$	
-2.96	Ca		Ox^{-2}			-2e			$CaOx$	
-2.96	Ca		$2Trop^-$			-2e			$Ca(Trop)_2$	
-2.95	Ba		$Tart^{-2}$			-2e			$BaTart$	
-2.95	Ca		$2Coy^-$			-2e			$Ca(Coy)_2$	
-2.95	Ra		$Aspa^{-2}$			-2e			$RaAspa$	
-2.95	Ra		Suc^{-2}			-2e			$RaSuc$	
-2.95	Sr		$S_2O_3^{-2}$			-2e			SrS_2O_3	
-2.94	Ba		Ap^{-2}			-2e			$BaAp$	
-2.94	Ba		Mal^{-2}			-2e			$BaMal$	
-2.94	Ca		HPO_4^{-2}			-2e			$CaHPO_4$	
-2.94	Ca		SO_4^{-2}			-2e			$CaSO_4$	
-2.94	Sr		$Tart^{-2}$			-2e			$SrTart$	
-2.93	Ba		$Aspa^{-2}$			-2e			$BaAspa$	
-2.93	Ba		Glu^-			-2e			$BaGlu^+$	
-2.93	Ba		IO_3^-			-2e			$BaIO_3^+$	
-2.93	Ba		NO_3^-			-2e			$BaNO_3^+$	
-2.93	Ba		Pht^{-2}			-2e			$BaPht$	
-2.93	Ba		Suc^{-2}			-2e			$BaSuc$	
-2.93	Ca		H_2Cit^-			-2e			CaH_2Cit^+	
-2.93	Ca		$S_2O_3^{-2}$			-2e			CaS_2O_3	
-2.93	H(g)		OH^-			-e			H_2O	
-2.93	K					-e	K^+			

EMF	Elem	Anion	Compound or anion	Cation	H_2O	e =	Cation	Anion	Complex or compound	Compound or element
-2.93	K		ClO_3^-			-e			$KClO_3$	
-2.93	Rb					-e	Rb^+			
-2.93	Sr		Ap^{-2}			-2e			SrAp	
-2.93	Sr		$Aspa^{-2}$			-2e			SrAspa	
-2.93	Sr		Mal^{-2}			-2e			SrMal	
-2.92	Ba		$Alan^-$			-2e			$BaAlan^+$	
-2.92	Ba		ClO_3^-			-2e			$BaClO_3^+$	
-2.92	Ba		Gl^-			-2e			$BaGl^+$	
-2.92	Ba		$Glac^-$			-2e			$BaGlac^+$	
-2.92	Ba		Gly^-			-2e			$BaGly^+$	
-2.92	Ba		OH^-			-2e			$BaOH^+$	
-2.92	Ba		Lac^-			-2e			$BaLac^+$	
-2.92	Ca		Ap^{-2}			-2e			CaAp	
-2.92	Ca		$Aspa^{-2}$			-2e			CaAspa	
-2.92	Ca		$Tart^{-2}$			-2e			CaTart	
-2.92	Cs					-e	Cs^+			
-2.92	K		NO_3^-			-e			KNO_3	
-2.92	Ra					-2e	Ra^{+2}			
-2.92	Sr		$Glac^-$			-2e			$SrGlac^+$	
-2.92	Sr		Glu^-			-2e			$SrGlu^+$	
-2.92	Sr		IO_3^-			-2e			$SrIO_3^+$	
-2.92	Sr		Suc^{-2}			-2e			SrSuc	
-2.91	Ba					-2e	Ba^{+2}			
-2.91	Ba		Ac^-			-2e			$BaAc^+$	
-2.91	Ba		Bu^-			-2e			$BaBu^+$	
-2.91	Ba		Pr^-			-2e			$BaPr^+$	
-2.91	Ca		$Alan^-$			-2e			$CaAlan^+$	
-2.91	Ca		Gl^-			-2e			$CaGl^+$	
-2.91	Ca		$Glgl^-$			-2e			$CaGlgl^+$	
-2.91	Ca		Glu^-			-2e			$CaGlu^+$	
-2.91	Ca		Mal^{-2}			-2e			CaMal	
-2.91	Ca		OH^-			-2e			$CaOH^+$	

EMF	Elem	Anion	Compound or anion	Cation	H_2O	e =	Cation	Anion	Complex or compound	Compound or element
−2.91	Na		$Amac^{-3}$			−e			$NaAmac^{-2}$	
−2.91	Sr		Gl^-			−2e			$SrGl^+$	
−2.91	Sr		Gly^-			−2e			$SrGly^+$	
−2.91	Sr		Lac^-			−2e			$SrLac^+$	
−2.91	Sr		NO_3^-			−2e			$SrNO_3^+$	
−2.91	Sr		OH^-			−2e			$SrOH^+$	
−2.90	Ca		$Glac^-$			−2e			$CaGlac^+$	
−2.90	Ca		Gly^-			−2e			$CaGly^+$	
−2.90	Ca		IO_3^-			−2e			$CaIO_3^+$	
−2.90	Ca		Pht^{-2}			−2e			$CaPht$	
−2.90	Ca		Suc^{-2}			−2e			$CaSuc$	
−2.90	K	BrO_3^-				−e				$KBrO_3$
−2.90	K		IO_3^-			−e			KIO_3	
−2.90	K	ClO_4^-				−e				$KClO_4$
−2.90	La		$3OH^-$			−3e			$La(OH)_3$	
−2.90	Sr		Ac^-			−2e			$SrAc^+$	
−2.90	Sr		Bu^-			−2e			$SrBu^+$	
−2.90	Sr		Pr^-			−2e			$SrPr^+$	
−2.89	Ca		Ac^-			−2e			$CaAc^+$	
−2.89	Ca		Bu^-			−2e			$CaBu^+$	
−2.89	Ca		Lac^-			−2e			$CaLac^+$	
−2.89	Ca		Pr^-			−2e			$CaPr^+$	
−2.89	Sr					−2e	Sr^{+2}			
−2.88	Ca		NO_3^-			−2e			$CaNO_3^+$	
−2.88	Ca		Ph			−2e			$CaPh^{+2}$	
−2.88	Sr		$2OH^-$			−2e			$Sr(OH)_2$	
−2.87	Ca					−2e	Ca^{+2}			
−2.87	Ca		Sal^{-2}			−2e			$CaSal$	
−2.87	Ce		$3OH^-$			−3e			$Ce(OH)_3$	
−2.86	Ca		Nac^-			−2e			$CaNac^+$	
−2.85	Na	$P_2O_7^{-4}$				−e			$Na(P_2O_7)^{-3}$	
−2.85	Pr		$3OH^-$			−3e			$Pr(OH)^{-3}$	

EMF	Elem	Anion	Compound or anion	Cation	H_2O	e =	Cation	Anion	Complex or compound	Compound or element
−2.84	La		Data^{-4}			−3e			LaData$^-$	
−2.84	Na		Nta^{-3}			−e			NaNta^{-2}	
−2.84	Nd		3OH$^-$			−3e			Nd(OH)$_3$	
−2.84	Pm		3OH$^-$			−3e			Pm(OH)$_3$	
−2.83	Eu		3OH$^-$			−3e			Eu(OH)$_3$	
−2.83	Sm		3OH$^-$			−3e			Sm(OH)$_3$	
−2.82	Ce		$P_2O_7^{-4}$			−3e			CeP$_2$O$_7^-$	
−2.82	Gd		3OH$^-$			−3e			Gd(OH)$_3$	
−2.81	Ba		2OH$^-$			−2e			Ba(OH)$_2$	
−2.81	Ce		Data^{-4}			−3e			CeData$^-$	
−2.81	La		Edta^{-4}			−3e			LaEdta$^-$	
−2.81	Na		Edta^{-4}			−e			NaEdta^{-3}	
−2.81	Y		3OH$^-$			−3e			Y(OH)$_3$	
−2.80	Pr		Data^{-4}			−3e			PrData$^-$	
−2.80	Yb					−2e	Yb^{+2}			
−2.79	Tb		3OH$^-$			−3e			Tb(OH)$_3$	
−2.78	Ce		Edta^{-4}			−3e			CeEdta$^-$	
−2.78			CuH			−e	H$^+$			Cu
−2.78	Dy		3OH$^-$			−3e			Dy(OH)$_3$	
−2.78	Eu		Data^{-4}			−3e			EuData$^-$	
−2.78	2La	3S^{-2}				−6e				La$_2$S$_3$
−2.78	Na		$P_3O_9^{-3}$			−e			NaP$_3$O$_9^{-2}$	
−2.78	Nd		Data^{-4}			−3e			NdData$^-$	
−2.77	Ce		OH$^-$			−3e			CeOH^{+2}	
−2.77	Gd		Data^{-4}			−3e			GdData$^-$	
−2.77	Ho		3OH$^-$			−3e			Ho(OH)$_3$	
−2.77	Na		Saa^{-3}			−e			NaSaa^{-2}	
−2.77	Pr		Edta^{-4}			−3e			PrEdta$^-$	
−2.77	Sm		Data^{-4}			−3e			SmData$^-$	
−2.77	Tb		Data^{-4}			−3e			TbData$^-$	
−2.76	Na		$P_4O_{12}^{-4}$			−e			NaP$_4$O$_{12}^{-3}$	
−2.75	Er		3OH$^-$			−3e			Er(OH)$_3$	

EMF	Elem	Anion	Compound or anion	Cation	H_2O	e	=	Cation	Anion	Complex or compound	Compound or element
−2.75	Na		$S_2O_3^{-2}$			−e				$NaS_2O_3^-$	
−2.75	Na		SO_4^{-2}			−e				$NaSO_4^-$	
−2.75	Nd		$Edta^{-4}$			−3e				$NdEdta^-$	
−2.75	Y		$Data^{-4}$			−3e				$YData^-$	
−2.74	Dy		$Data^{-4}$			−3e				$DyData^-$	
−2.74	Eu		$Edta^{-4}$			−3e				$EuEdta^-$	
−2.74	Sm		$Edta^{-4}$			−3e				$SmEdta^-$	
−2.74	Tm		$3OH^-$			−3e				$Tm(OH)_3$	
−2.73	Gd		$Edta^{-4}$			−3e				$GdEdta^-$	
−2.73	Tb		$Edta^{-4}$			−3e				$TbEdta^-$	
−2.73	Yb		$3OH^-$			−3e				$Yb(OH)_3$	
−2.72	La		$2Amac^{-3}$			−3e				$La(Amac)_2^{-3}$	
−2.72	La		Nta^{-3}			−3e				$LaNta$	
−2.72	Lu		$3OH^-$			−3e				$Lu(OH)_3$	
−2.71	Er		$Data^{-4}$			−3e				$ErData^-$	
−2.71	Na					−e		Na^+			
−2.71	Y		$Edta^{-4}$			−3e				$YEdta^-$	
−2.70	Cm					−3e		Cm^{+3}			
−2.70	Dy		$Edta^{-4}$			−3e				$DyEdta^-$	
−2.69	2Ce	$3S^{-2}$				−6e					Ce_2S_3
−2.69	La		$Oxin^-$			−3e				$LaOxin^{+2}$	
−2.69	Mg		$2OH^-$			−2e				$Mg(OH)_2$	
−2.69	Na	BrO_3^-				−e					$NaBrO_3$
−2.69	Yb		$Data^{-4}$			−3e				$YbData^-$	
−2.68	Ce		$2Amac^{-3}$			−3e				$Ce(Amac)_2^{-3}$	
−2.68	Lu		$Data^{-4}$			−3e				$LuData^-$	
−2.68	Na		OH^-			−e				$NaOH$	
−2.67	Ho		$Edta^{-4}$			−3e				$HoEdta^-$	
−2.67	La		$2Nta^{-3}$			−3e				$La(Nta)_2^{-3}$	
−2.67	La		$2Oxin^-$			−3e				$La(Oxin)_2^+$	
−2.66	Ce		$Oxin^-$			−3e				$CeOxin^{+2}$	
−2.65	Er		$Edta^{-4}$			−3e				$ErEdta^-$	

EMF	Elem	Anion	Compound or anion	Cation	H_2O	e =	Cation	Anion	Complex or compound	Compound or element
-2.65	La		$P_4O_{12}^{-4}$			-3e			$LaP_4O_{12}^{-}$	
-2.64	Ce		Nta^{-3}			-3e			CeNta	
-2.64	Ce		$2Oxin^{-}$			-3e			$Ce(Oxin)_2^{+}$	
-2.64	Lu		$Edta^{-4}$			-3e			$LuEdta^{-}$	
-2.64	Yb		$Edta^{-4}$			-3e			$YbEdta^{-}$	
-2.63	2Be		$6OH^{-}$			-4e	$Be_2O_3^{-2}$			$3H_2O$
-2.63	Ce		Oxd^{-}			-3e			$CeOxd^{+2}$	
-2.63	La		$P_3O_9^{-3}$			-3e			LaP_3O_9	
-2.62	La		$AcAc^{-}$			-3e			$LaAcAc^{+2}$	
-2.62	La		OH^{-}			-3e			$LaOH^{+2}$	
-2.62	La		$Oxac^{-2}$			-3e			$LaOxac^{+}$	
-2.61	Be		$2OH^{-}$			-2e			H_2O	BeO
-2.61	Ce		Ox^{-2}			-3e			$CeOx^{+}$	
-2.61	Sc		$3OH^{-}$			-3e			$Sc(OH)_3$	
-2.60	Ac					-3e	Ac^{+3}			
-2.60	La		$(CoCy)^{-3}$			-3e			$La(CoCy)$	
-2.60	Mg		$Amac^{-3}$			-2e			$MgAmac^{-}$	
-2.60	La		$(FeCy)^{-3}$			-3e			$La(FeCy)$	
-2.60	Mg		$Edta^{-4}$			-2e			$MgEdta^{-2}$	
-2.59	La		$2AcAc^{-}$			-3e			$La(AcAc)_2^{+}$	
-2.59	La		SO_4^{-2}			-3e			$LaSO_4^{+}$	
-2.58	Ce		$AcAc^{-}$			-3e			$CeAcAc^{+2}$	
-2.58	La		$3AcAc^{-}$			-3e			$La(AcAc)_3$	
-2.58	U		OH^{-}			-4e			UOH^{+3}	
-2.57	La		F^{-}			-3e			LaF^{+2}	
-2.57	Nd		Ox^{-2}			-3e			$NdOx^{+}$	
-2.56	Ce		$2AcAc^{-}$			-3e			$Ce(AcAc)_2^{+}$	
-2.56	Ce		$2Ox^{-2}$			-3e			$Ce(Ox)_2^{-}$	
-2.56	Mg		Ers^{-3}			-2e			$MgErs^{-}$	
-2.56	Mg		Esb^{-3}			-2e			$MgEsb^{-}$	
-2.56	Mg		Nta^{-3}			-2e			$MgNta^{-}$	
-2.55	Ce		SO_4^{-2}			-3e			$CeSO_4^{+}$	

EMF	Elem	Anion	Compound or anion	Cation	H_2O	e =	Cation	Anion	Complex or compound	Compound or element
-2.55	Mg		Esa^{-3}			-2e			$MgEsa^-$	
-2.55	Mg		Est^{-3}			-2e			$MgEst^-$	
-2.54	Ce		F^-			-3e			CeF^{+2}	
-2.54	Nd		$AcAc^-$			-3e			$NdAcAc^{+2}$	
-2.53	Eu		$AcAc^-$			-3e			$EuAcAc^{+2}$	
-2.53	Mg		$Oxin^-$			-2e			$MgOxin^+$	
-2.53	Pr		SO_4^{-2}			-3e			$PrSO_4^+$	
-2.53	Sm		$AcAc^-$			-3e			$SmAcAc^{+2}$	
-2.52	Ce		SO_4^{-2}			-3e			$CeSO_4^+$	
-2.52	La					-3e	La^{+3}			
-2.52	Mg		$Ist(a)^-$			-2e			$Mg(Ist(a))^+$	
-2.52	Mg		$Ist(b)^-$			-2e			$Mg(Ist(b))^+$	
-2.52	Mg		$Met(a)^-$			-2e			$Mg(Met(a))^+$	
-2.52	Mg		$Met(b)^-$			-2e			$Mg(Met(b))^+$	
-2.52	Mg		$Tmta^{-4}$			-2e			$MgTmta^{-2}$	
-2.51	Ce		Ac^-			-3e			$CeAc^{+2}$	
-2.51	Gd		$Oxac^{-2}$			-3e			$GdOxac^+$	
-2.51	Mg		$P_2O_7^{-4}$			-2e			$MgP_2O_7^{-2}$	
-2.51	Nd		$2Ox^{-2}$			-3e			$Nd(Ox)_2^-$	
-2.51	U		Tf^-			-4e			Utf^{+3}	
-2.50	Ce		$2Ac^-$			-3e			$CeAc_2^+$	
-2.50	Ce		$3Ox^{-2}$			-3e			$Ce(Ox)_3^{-3}$	
-2.50	Eu		$2AcAc^-$			-3e			$Eu(AcAc)_2^+$	
-2.50	Hf		$4OH^-$			-4e			$HfO(OH)_2$	H_2O
-2.50	Mg		$2Oxim^-$			-2e			$Mg(Oxim)_2$	
-2.50	Mg		$Trop^-$			-2e			$MgTrop^+$	
-2.50	Nd		SO_4^{-2}			-3e			$NdSO_4^+$	
-2.50	No					-3e	No^{+3}			
-2.50	Sm		$2AcAc^-$			-3e			$Sm(AcAc)_2^+$	
-2.50	Y		$AcAc^-$			-3e			$YAcAc^{+2}$	
-2.49	Ce		$3Ac^-$			-3e			$CeAc_3$	
-2.49	Gd		$2Oxac^{-2}$			-3e			$Gd(Oxac)_2^-$	

EMF	Elem	Anion	Compound or anion	Cation	H_2O	e	=	Cation	Anion	Complex or compound	Compound or element
-2.49	Ce		Br^-			-3e					$CeBr^{+2}$
-2.49	Mg		$2Ist(a)^-$			-2e					$Mg(Ist(a))_2$
-2.49	Mg		$MeOxin^-$			-2e					$MgMeOxin^+$
-2.49	Mg		$Ndap^{-3}$			-2e					$MgNdap^-$
-2.49	Mg		$mOxin^-$			-2e					$Mg(mOxin)^+$
-2.49	Mg		$P_4O_{12}^{-4}$			-2e					$MgP_4O_{12}^{-2}$
-2.49	Nd		$3AcAc^-$			-3e					$Nd(AcAc)_3$
-2.48	Ce					-3e		Ce^{+3}			
-2.48	Eu		$3AcAc^-$			-3e					$Eu(AcAc)_3$
-2.48	Mg		$(FeCy)^{-4}$			-2e					$Mg(FeCy)^{-2}$
-2.48	Mg		$2Ist(b)^-$			-2e					$Mg(Ist(b))_2$
-2.48	Mg		$2Met(a)^-$			-2e					$Mg(Met(a))_2$
-2.48	Mg		$2Met(b)^-$			-2e					$Mg(Met(b))_2$
-2.48	Mg		$meOxin^-$			-2e					$Mg(meOxin)^+$
-2.48	Sm		SO_4^{-2}			-3e					$SmSO_4^+$
-2.48	Th		$4OH^-$			-4e					$Th(OH)_4$
-2.48	Y		$Oxac^{-2}$			-3e					$YOxac^+$
-2.47	Gd		SO_4^{-2}			-3e					$GdSO_4^+$
-2.47	Mg		$2MeOxin^-$			-2e					$Mg(MeOxin)_2$
-2.47	Mg		$2mOxin^-$			-2e					$Mg(mOxin)_2$
-2.47	Mg		$2Trop^-$			-2e					$Mg(Trop)_2$
-2.47	Sm		$3AcAc^-$			-3e					$Sm(AcAc)_3$
-2.46	Dy		$Oxac^{-2}$			-3e					$DyOxac^+$
-2.46	Mg		Has^-			-2e					$MgHas^+$
-2.46	Mg		$2meOxin^-$			-2e					$Mg(meOxin)_2$
-2.46	Pr					-3e		Pr^{+3}			
-2.46	Y		$2AcAc^-$			-3e					$Y(AcAc)_2^+$
-2.45	Ce		$3AcAc^-$			-3e					$Ce(AcAc)_3$
-2.45	Gd		F^-			-3e					GdF^{+2}
-2.45	Mg		$AcAc^-$			-2e					$MgAcAc^+$
-2.45	Mg		$DmHas^-$			-2e					$Mg(DmHas)^+$
-2.45	Mg		$Imda^{-2}$			-2e					$MgImda$

EMF	Elem	Anion	Compound or anion	Cation	H_2O	e =	Cation	Anion	Complex or compound	Compound or element
-2.45	Mg		$Mcin^-$			-2e			$MgMcin^+$	
-2.45	Mg		$Ndpa^{-3}$			-2e			$MgNdpa^-$	
-2.45	Mg		Oxd^-			-2e			$MgOxd^+$	
-2.45	U		SO_4^{-2}			-4e			USO_4^{+2}	
-2.45	Y		$2Oxac^{-2}$			-3e			$Y(Oxac)_2^-$	
-2.44	Dy		$2Oxac^{-2}$			-3e			$Dy(Oxac)_2^-$	
-2.44	Mg		Gl^-			-2e			$MgGl^+$	
-2.44	Mg	PO_4^{-3}				-2e			$MgPO_4^-$	
-2.44	Mg		$Himda^{-2}$			-2e			$MgHimda$	
-2.44	Mg		Hox^-			-2e			$MgHox^+$	
-2.44	Mg		Ox^{-2}			-2e			$MgOx$	
-2.44	Mg		$P_3O_9^{-3}$			-2e			$MgP_3O_9^-$	
-2.44	Y		SO_4^{-2}			-3e			YSO_4^+	
-2.43	Mg		$2Amac^{-3}$			-2e			$Mg(Amac)_2^{-4}$	
-2.43	Mg		Cin^-			-2e			$MgCin^+$	
-2.43	Mg		Cit^{-3}			-2e			$MgCit^-$	
-2.43	Mg		$2DmHas^-$			-2e			$Mg(DmHas)_2$	
-2.43	Mg		$2Has^=$			-2e			$Mg(Has)_2$	
-2.43	Mg		$2Hox^-$			-2e			$Mg(Hox)_2$	
-2.43	Mg		$3Ist(b)^-$			-2e			$Mg(Ist(b))_3^-$	
-2.43	Mg		$3Met(b)^-$			-2e			$Mg(Met(b))_3^-$	
-2.43	Mg		$2Nta^{-3}$			-2e			$Mg(Nta)_2^{-4}$	
-2.43	Mg		$2Oxd^-$			-2e			$Mg(Oxd)_2$	
-2.43	Nd					-3e	Nd^{+3}			
-2.43	Sc	$3F^-$				-3e				ScF_3
-2.43	U		$2SO_4^{-2}$			-4e			$U(SO_4)_2$	
-2.43	Y		$3AcAc^-$			-3e			$Y(AcAc)_3$	
-2.42			CdH			-e	H^+			Cd
-2.42	Mg		$2AcAc^-$			-2e			$Mg(AcAc)_2$	
-2.42	Mg		Mal^{-2}			-2e			$MgMal$	
-2.42	Mg		$2Mcin^-$			-2e			$Mg(Mcin)_2$	
-2.42	Mg		$3Met(a)^-$			-2e			$Mg(Met(a))_3^-$	

EMF	Elem	Anion	Compound or anion	Cation	H_2O	e =	Cation	Anion	Complex or compound	Compound or element
-2.42	Mg		OH^-			-2e			$MgOH^+$	
-2.42	Mg		Saa^{-3}			-2e			$MgSaa^-$	
-2.42	Pm					-3e	Pm^{+3}			
-2.42	Pu		$3OH^-$			-3e			$Pu(OH)_3$	
-2.42	U		CNS^-			-4e			$UCNS^{+3}$	
-2.41	Eu					-3e	Eu^{+3}			
-2.41	Mg		$Aspa^{-2}$			-2e			$MgAspa$	
-2.41	Mg		HPO_4^{-2}			-2e			$MgHPO_4$	
-2.41	Mg		SO_4^{-2}			-2e			$MgSO_4$	
-2.41	Sm					-3e	Sm^{+3}			
-2.41	U		Cl^-			-4e			UCl^{+3}	
-2.41	U		$2CNS^-$			-4e			$U(CNS)_2^{+2}$	
-2.41	Yb		Ox^{-2}			-3e			$YbOx^+$	
-2.40	Bk					-3e	Bk^{+3}			
-2.40	Gd					-3e	Gd^{+3}			
-2.40	Mg		$Alan^-$			-2e			$MgAlan^+$	
-2.40	Mg		$2Cin^-$			-2e			$Mg(Cin)_2$	
-2.40	U		Br^-			-4e			UBr^{+3}	
-2.39	Ho		SO_4^{-2}			-3e			$HoSO_4^+$	
-2.39	Mg		Ap^{-2}			-2e			$MgAp$	
-2.39	Mg		$S_2O_3^{-2}$			-2e			MgS_2O_3	
-2.39	Tb					-3e	Tb^{+3}			
-2.39	U		$4OH^-$			-4e			$2H_2O$	UO_2
-2.38	Lu		$Oxac^{-2}$			-3e			$LuOxac^+$	
-2.38	Mg		F^-			-2e			MgF^+	
-2.38	Mg		Ph			-2e			$MgPh^{+2}$	
-2.38	Mg		Suc^{-2}			-2e			$MgSuc$	
-2.38	Mg		$Tart^{-2}$			-2e			$MgTart$	
-2.37	Er		SO_4^{-2}			-3e			$ErSO_4^+$	
-2.37	Mg		Dge^-			-2e			$MgDge^+$	
-2.37	Mg		$Glac^-$			-2e			$MgGlac^+$	
-2.37	Mg		$Glgl^-$			-2e			$MgGlgl^+$	

EMF	Elem	Anion	Compound or anion	Cation	H_2O	e =	Cation	Anion	Complex or compound	Compound or element
−2.37	Mg		Gly⁻			−2e			MgGly⁺	
−2.37	Mg		Lac⁻			−2e			MgLac⁺	
−2.37	Y					−3e	Y⁺³			
−2.36	Mg					−2e	Mg⁺²			
−2.36	Mg		Ac⁻			−2e			MgAc⁺	
−2.36	Mg		Bu⁻			−2e			MgBu⁺	
−2.36	Mg		Glu⁻			−2e			MgGlu⁺	
−2.36	Mg		IO_3^-			−2e			$MgIO_3^+$	
−2.36	Mg		Pr⁻			−2e			MgPr⁺	
−2.36	Yb		2Ox⁻²			−3e			$Yb(Ox)_2^-$	
−2.36	Zr		4OH⁻			−4e			H_2ZrO_3	H_2O
−2.35	Dy					−3e	Dy⁺³			
−2.35	Lu		2Oxac⁻²			−3e			$Lu(Oxac)_2^-$	
−2.35	Mg		Dyp			−2e			MgDyp⁺²	
−2.35	Mg		NH_3			−2e			$Mg(NH_3)^{+2}$	
−2.34	Mg		$2NH_3$			−2e			$Mg(NH_3)_2^{+2}$	
−2.34	Mg		NO_3^-			−2e			$MgNO_3^+$	
−2.34	Yb		SO_4^{-2}			−3e			$YbSO_4^+$	
−2.33	Al		4OH⁻			−3e		$H_2AlO_3^-$		H_2O
−2.33	Mg		Nac⁻			−2e			MgNac⁺	
−2.33	Mg		$3NH_3$			−2e			$Mg(NH_3)_3^{+2}$	
−2.33	Sc	2F⁻				−3e			ScF_2^+	
−2.32	Am					−3e	Am⁺³			
−2.32	Ho					−3e	Ho⁺³			
−2.32	Mg		$4NH_3$			−2e			$Mg(NH_3)_4^{+2}$	
−2.32	Sc		OH⁻			−3e			ScOH⁺²	
−2.31	Mg		$5NH_3$			−2e			$Mg(NH_3)_5^{+2}$	
−2.30	Al		3OH⁻			−3e			$Al(OH)_3$	
−2.30	Er					−3e	Er⁺³			
−2.30	Mg		$6NH_3$			−2e			$Mg(NH_3)_6^{+2}$	
−2.28			HgH			−e	H⁺			Hg
−2.28	Tm					−3e	Tm⁺³			

EMF	Elem	Anion	Compound or anion	Cation	H_2O	e =	Cation	Anion	Complex or compound	Compound or element
-2.27	Yb					-3e	Yb^{+3}			
-2.26	Lu					-3e	Lu^{+3}			
-2.25		H^-				-e				$\frac{1}{2}H_2$
-2.24	Sc		$AcAc^-$			-3e			$ScAcAc^{+2}$	
-2.24	Sc	F^-				-3e			ScF^{+3}	
-2.22	Sc		$2AcAc^-$			-3e			$Sc(AcAc)_2^+$	
-2.20		OH^-	$U(OH)_3$			-e			$U(OH)_4$	
-2.17	U		$3OH^-$			-3e			$U(OH)_3$	
-2.11	H(g)					-e	H^+			
-2.10	Cf					-3e	Cf^{+3}			
-2.10	Fm					-3e	Fm^{+3}			
-2.08	Sc					-3e	Sc^{+3}			
-2.07	Al		$6F^-$			-3e			AlF_6^{-3}	
-2.05	P		$2OH^-$			-e		$H_2PO_2^-$		
-2.03	Be		$Met(a)^-$			-2e			$Be(Met(a))^+$	
-2.03	Pu					-3e	Pu^{+3}			
-2.03	Th		$AcAc^-$			-4e			$ThAcAc^{+3}$	
-2.02	Al		$Data^{-4}$			-3e			$AlData^-$	
-2.02	Be		$Ist(a)^-$			-2e			$Be(Ist(a))^+$	
-2.01	2Pm				$3H_2O$	-6e	$6H^+$			Pm_2O_3
-2.01	Th		$2AcAc^-$			-4e			$Th(AcAc)_2^{+2}$	
-2.01	Th		F^-			-4e			ThF^{+3}	
-2.01	Th		$3IO_3^-$			-4e			$Th(IO_3)_3^+$	
-2.01	Th		Tf^-			-4e			$ThTf^{+3}$	
-2.00	2Eu				$3H_2O$	-6e	$6H^+$			Eu_2O_3
-2.00	2Sm				$3H_2O$	-6e	$6H^+$			Sm_2O_3
-2.00	2Tb				$3H_2O$	-6e	$6H^+$			Tb_2O_3
-1.99	Th		$3AcAc^-$			-4e			$Th(AcAc)_3^+$	
-1.99	2Gd				$3H_2O$	-6e				Gd_2O_3
-1.99	Th		$2F^-$			-4e			ThF_2^{+2}	
-1.98	Be		$Met(b)^-$			-2e			$Be(Met(b))^+$	
-1.98	Y				$3H_2O$	-3e	$3H^+$			$Y(OH)_3$

EMF	Elem	Anion	Compound or anion	Cation	H_2O	e =	Cation	Anion	Complex or compound	Compound or element
-1.97	Be		$2Ist(a)^-$			-2e			$Be(Ist(a))_2$	
-1.97	Be		$Ist(b)^-$			-2e			$Be(Ist(b))^+$	
-1.97	Be		$2Met(a)^-$			-2e			$Be(Met(a))_2$	
-1.97	Sr				H_2O	-2e	$2H^+$			SrO
-1.97	Th		$3F^-$			-4e			ThF_3^+	
-1.97	Th		$2IO_3^-$			-4e			$Th(IO_3)_2^{+2}$	
-1.96	2Dy				$3H_2O$	-6e	$6H^+$			Dy_2O_3
-1.96	Th		$4AcAc^-$			-4e			$Th(AcAc)_4$	
-1.96	Th		$H_2PO_4^-$			-4e			$ThH_2PO_4^{+3}$	
-1.96	Th		$2H_2PO_4^-$			-4e			$Th(H_2PO_4)_2^{+2}$	
-1.95	Al		Sal^{-2}			-3e			$AlSal^+$	
-1.95			InH			-e	H^+			In
-1.95	H	NH_2^-				-e				NH_3
-1.95	Be		$Trop^-$			-2e			$BeTrop^+$	
-1.95	Pa					-3e	Pa^{+3}			
-1.95	Th		SO_4^{-2}			-4e			$ThSO_4^{+2}$	
-1.94	2Ho				$3H_2O$	-6e	$6H^+$			Ho_2O_3
-1.94	Th		IO_3^-			-4e			$ThIO_3^{+3}$	
-1.93	Be		$AcAc^-$			-2e			$BeAcAc^+$	
-1.93	Be		$2Met(b)^-$			-2e			$Be(Met(b))_2$	
-1.93	Th		H_3PO_4			-4e			$ThH_3PO_4^{+4}$	
-1.93	Th		$2SO_4^{-2}$			-4e			$Th(SO_4)_2$	
-1.92	Be		$2Ist(b)^-$			-2e			$Be(Ist(b))_2$	
-1.92	Be		OH^-			-2e			$BeOH^+$	
-1.92	2Er				$3H_2O$	-6e	$6H^+$			Er_2O_3
-1.91	Be		$2Trop^-$			-2e			$Be(Trop)_2$	
-1.91	Th		BrO_3^-			-4e			$ThBrO_3^{+3}$	
-1.91	Th		NO_3^-			-4e			$ThNO_3^{+3}$	
-1.91	Th		OH^-			-4e			$ThOH^{+3}$	
-1.91	2Tm				$3H_2O$	-6e	$6H^+$			Tm_2O_3
-1.90	Be		$2AcAc^-$			-2e			$Be(AcAc)_2$	
-1.90	Ca				H_2O	-2e	$2H^+$			CaO

EMF	Elem	Anion	Compound or anion	Cation	H_2O	e =	Cation	Anion	Complex or compound	Compound or element
-1.90	Th					-4e	Th^{+4}			
-1.90	Th		$2BrO_3^-$			-4e			$Th(BrO_3)_2^{+2}$	
-1.90	Th		Cl^-			-4e			$ThCl^{+3}$	
-1.90	Th		ClO_3^-			-4e			$ThClO_3^{+3}$	
-1.90	2Yb				$3H_2O$	-6e	$6H^+$			Yb_2O_3
-1.89	Th		$3Cl^-$			-4e			$ThCl_3^+$	
-1.89	2Lu				$3H_2O$	-6e	$6H^+$			Lu_2O_3
-1.89	Th		$4Cl^-$			-4e			$ThCl_4$	
-1.88	Am				$3H_2O$	-3e	$3H^+$			$Am(OH)_3$
-1.88	Th		$2Cl^-$			-4e			$ThCl_2^{+2}$	
-1.87			TlH			-e	H^+			Tl
-1.86	Np					-3e	Np^{+3}			
-1.86	2La				$3H_2O$	-6e	$6H^+$			La_2O_3
-1.85	Be					-2e	Be^{+2}			
-1.84	Al		$AcAc^-$			-3e			$AlAcAc^{+2}$	
-1.84	Al		OH^-			-3e			$AlOH^{+2}$	
-1.83	$2P_2$				$3H_2O$	-6e	$6H^+$			Pr_2O_3
-1.83	Al		$2AcAc^-$			-3e			$Al(AcAc)_2^+$	
-1.83	Be		Cit^{-3}			-2e			$BeCit^-$	
-1.83	Be		F^-			-2e			BeF^+	
-1.81	2Nd				$3H_2O$	-6e	$6H^+$			Nd_2O_3
-1.80	Al	Cl^-				-3e			$AlCl^{+2}$	
-1.79	Al		F^-			-3e			AlF^{+2}	
-1.79	B		$4OH^-$			-3e		$H_2BO_3^-$	H_2O	
-1.79	Be				H_2O	-2e	$2H^+$			BeO
-1.79	U					-3e	U^{+3}			
-1.78	Al		$3AcAc^-$			-3e			$Al(AcAc)_3$	
-1.77	Al		$2F^-$			-3e			AlF_2^+	
-1.77	Be		$HCit^{-2}$			-2e			BeHCit	
-1.76	Be		$2F^-$			-2e			BeF_2	
-1.75	Al		$3F^-$			-3e			AlF_3	
-1.75	Np		SO_4^{-2}			-4e			$NpSO_4^{+2}$	

EMF	Elem	Anion	Compound or anion	Cation	H_2O	e =	Cation	Anion	Complex or compound	Compound or element
-1.74	Al		$3Ox^{-2}$			-3e			$Al(Ox)_3^{-3}$	
-1.74	Be		H_2Cit^-			-2e			BeH_2Cit^+	
-1.73	Np		$2SO_4^{-2}$			-4e			$Np(SO_4)_2$	
-1.72	Al		$4F^-$			-3e			AlF_4^-	
-1.71	Be		Nac^-			-2e			$BeNac^+$	
-1.70	Al		$5F^-$			-3e			AlF_5^{-2}	
-1.70	Hf					-4e	Hf^{+4}			
-1.70	Pa					-4e	Pa^{+4}			
-1.70	Si		$6OH^-$			-4e		SiO_3^{-2}	$3H_2O$	
-1.68	Al		$6F^-$			-3e			AlF_6^{-3}	
-1.68	2Am				$3H_2O$	-6e	$6H^+$			Am_2O_3
-1.68	Al		Nac^-			-3e			$AlNac^{+2}$	
-1.68	2Y				$3H_2O$	-6e	$6H^+$			Y_2O_3
-1.67	V		OH^-			-3e	VOH^{+2}			
-1.66	Al					-3e	Al^{+3}			
-1.66	V		$2OH^-$			-3e	$V(OH)_2^+$			
-1.66	Zr		F^-			-4e			ZrF^{+3}	
-1.64	Zr		$2F^-$			-4e			ZrF_2^{+2}	
-1.63	Ti					-2e	Ti^{+2}			
-1.62		$4OH^-$	$U(OH)_4$	$2Na^+$			-2e			Na_2UO_4	$4H_2O$
-1.62	Zr		$3F^-$			-4e			ZrF_3^+	
-1.59	2Sc				$3H_2O$	-6e	$6H^+$			Sc_2O_3
-1.59	Zr		SO_4^{-2}			-4e			$ZrSO_4^{+2}$	
-1.57		$H_2PO_2^-$	$3OH^-$			-2e		HPO_3^{-2}	$2H_2O$	
-1.57	Zr		$2SO_4^{-2}$			-4e			$Zr(SO_4)_2$	
-1.55	Mn		$Data^{-4}$			-2e			$MnData^{-2}$	
-1.55	Mn		$2OH^-$			-2e			$Mn(OH)_2$	
-1.55	Zr				$4H_2O$	-4e	$4H^+$			$Zr(OH)_4$
-1.55	Zr		$3SO_4^{-2}$			-4e			$Zr(SO_4)_3^{-2}$	
-1.54	2Al				$6H_2O$	-6e	$6H^+$		$3H_2O$	Al_2O_3
-1.53	Zr					-4e	Zr^{+4}			
-1.51	Hf				$2H_2O$	-4e	$4H^+$			HfO_2

EMF	Elem	Anion	Compound or anion	Cation	H_2O	e =	Cation	Anion	Complex or compound	Compound or element
-1.51			PbH_2			-2e	$2H^+$			Pb
-1.50	Mn		CO_3^{-2}			-2e			$MnCO_3(c)$	
-1.49	V		CNS^-			-3e			$VCNS^{+2}$	
-1.48	Cr		$3OH^-$			-3e			$Cr(OH)_3(c)$	
-1.48	Mn		CO_3^{-2}			-2e			$MnCO_3(ppt)$	
-1.46	Zr				$2H_2O$	-4e	$4H^+$			ZrO_2
-1.45	Mn		$Edta^{-4}$			-2e			$MnEdta^{-2}$	
-1.42	Np				$3H_2O$	-3e	$3H^+$			$Np(OH)_3$
-1.41	Zn		S^{-2}			-2e			ZnS	
-1.37	Mn		Hed^{-3}			-2e			$MnHed^-$	
-1.34	Cr		$3OH^-$			-3e			$Cr(OH)_3$	
-1.32	Ra				H_2O	-2e	$2H^+$			RaO
-1.31	Zn		$Data^{-4}$			-2e			$ZnData^{-2}$	
-1.30			CsH			-2e	Cs^+			H^+
-1.30			RbH			-2e	Rb^+			H^+
-1.29	Mn		$Oxim^-$			-2e			$MnOxim^+$	
-1.28	Mn		Oxd^-			-2e			$MnOxd^+$	
-1.28	Cu	$4CN^-$				-e	$Cu(CN)_4^{-3}$			
-1.27	Cr		$4OH^-$			-3e			CrO_2	$2H_2O$
-1.27			KH			-2e	K^+			H^+
-1.27	Mn		Nta^{-3}			-2e			$MnNta^-$	
-1.26	Al				$2H_2O$	-3e	$4H^+$	AlO_2^-		
-1.26	Mn		$2Oxim^-$			-2e			$Mn(Oxim)_2$	
-1.26	Zn		$4CN^-$			-2e			$Zn(CN)_4^{-2}$	
-1.25	Mn		$2Oxd^-$			-2e			$Mn(Oxd)_2$	
-1.25	Zn		$Edta^{-4}$			-2e			$ZnEdta^{-2}$	
-1.25	Zn		$2OH^-$			-2e			$Zn(OH)_2$	
-1.24	Si		$6F^-$			-4e			SiF_6^{-2}	
-1.22	Ga		$4OH^-$			-3e		$H_2GaO_3^-$	H_2O	
-1.22	Mn		$Himda^{-2}$			-2e			MnHimda	
-1.22	Mn		Tate			-2e			$Mn(Tate)^{+2}$	
-1.22	Zn		$4OH^-$			-2e		ZnO_2^{-2}	$2H_2O$	

EMF	Elem	Anion	Compound or anion	Cation	H_2O	e =	Cation	Anion	Complex or compound	Compound or element
-1.21	Mn		$P_4O_{12}^{-4}$			-2e			$MnP_4O_{12}^{-2}$	
-1.21				Yb^{+2}		-e	Yb^{+3}			
-1.19	Mn	.	Teta			-2e			$Mn(Teta)^{+2}$	
-1.19	Pu		OH^-			-4e			$PuOH^{+3}$	
-1.19	Ti		$6F^-$			-4e			TiF_6^{-2}	
-1.19	V					-2e	V^{+2}			
-1.19	Zn		Hed^{-3}			-2e			$ZnHed^-$	
-1.19	Zn		Tate			-2e			$Zn(Tate)^{+2}$	
-1.18	Cd		S^{-2}			-2e			CdS	
-1.18	Mn					-2e	Mn^{+2}			
-1.18	Mn		$AcAc^-$			-2e			$MnAcAc^+$	
-1.18	Pu		Tf^-			-4e			$PuTf^{+3}$	
-1.17	Cu	$3CN^-$				-e			$Cu(CN)_3^{-2}$	
-1.17	Mn		$2Amac^{-3}$			-2e			$Mn(Amac)_2^{-4}$	
-1.17	Mn		$Aspa^{-2}$			-2e			MnAspa	
-1.17	Mn		Deta			-2e			$MnDeta^{+2}$	
-1 17	Mn		$2Himda^{-2}$			-2e			$Mn(Himda)_2^{-2}$	
-1.17	Mn		Ox^{-2}			-2e			MnOx	
-1.17	Tl		OH^-			-e			TlOH	
-1.16	Fe	$6CN^-$				-2e		$FeCy^{-4}$		
-1.16	Mn		$2Nta^{-3}$			-2e			$Mn(Nta)_2^{-4}$	
-1.16			LiH			-2e	H^+			Li^+
-1.16	Mn		$P_3O_9^{-3}$			-2e			$MnP_3O_9^-$	
-1.16	Mn		$Sald^-$			-2e			$MnSald^+$	
-1.16			NaH			-2e	H^+			Na^+
-1.16	Pu		F^-			-4e			PuF^{+3}	
-1.15	Mn		Gl^-			-2e			$MnGl^+$	
-1.15	Mn		Mal^{-2}			-2e			MnMal	
-1.15	Mn		OH^-			-2e			$MnOH^+$	
-1.15				Sm^{+2}		-e	Sm^{+3}			
-1.15	6V		$330H^-$			-30e			$HV_6O_{17}^{-3}$	$16H_2O$
-1.14	Mn		$2AcAc^-$			-2e			$Mn(AcAc)_2$	

EMF	Elem	Anion	Compound or anion	Cation	H_2O	e =	Cation	Anion	Complex or compound	Compound or element
−1.14	Mn		$Alan^-$			−2e			$MnAlan^+$	
−1.14	Mn		$2Alan^-$			−2e			$Mn(Alan)_2$	
−1.14	Mn		Dge^-			−2e			$MnDge^+$	
−1.14	Mn		$2Sald^-$			−2e			$Mn(Sald)_2$	
−1.14		Te^{-2}				−2e				Te
−1.13	Mn		2Deta			−2e			$Mn(Deta)_2^{+2}$	
−1.13	Mn		En			−2e			$MnEn^{+2}$	
−1.12	Mn		$2Dge^-$			−2e			$Mn(Dge)_2$	
−1.12	Mn		Dyp			−2e			$MnDyp^{+2}$	
−1.12	Mn		SO_4^{-2}			−2e			$MnSO_4$	
−1.12		HPO_3^{-2}	$3OH^-$			−2e		PO_4^{-3}	$2H_2O$	
−1.12		$S_2O_4^{-2}$	$4OH^-$			−2e		$2SO_3^{-2}$	$2H_2O$	
−1.12	Zn		Teta			−2e			$ZnTeta^{+2}$	
−1.11			BaH_2			−4e	$2H^+$				Ba^{+2}
−1.11	Mn		2En			−2e			$Mn(En)_2^{+2}$	
−1.11	Mn		$Glgl^-$			−2e			$MnGlgl^+$	
−1.11	Mn		$S_2O_3^{-2}$			−2e			MnS_2O_3	
−1.11	Pu		SO_4^{-2}			−4e			$PuSO_4^{+2}$	
−1.10	Nb					−3e	Nb^{+3}			
−1.09	Mn		Ac^-			−2e			$MnAc^+$	
−1.09			SrH_2			−4e	$2H^+$				Sr^{+2}
−1.08	Mn		3En			−2e			$Mn(En)_3^{+2}$	
−1.07	Mn		NH_3			−2e			$MnNH_3^{+2}$	
−1.07	Pu		NO_3^-			−4e			$PuNO_3^{+3}$	
−1.07			SnH_4			−4e	$4H^+$				Sn
−1.07	Zn		Nta^{-3}			−2e			$ZnNta^-$	
−1.06	Mn		$2NH_3$			−2e			$Mn(NH_3)_2^{+2}$	
−1.06	2Tl	CrO_4^{-2}				−2e				Tl_2CrO_4
−1.06	Zn		CO_3^{-2}			−2e			$ZnCO_3$	
−1.06	Zn		$Ndap^{-3}$			−2e			$ZnNdap^-$	
−1.05	Mo		$8OH^-$			−6e		MoO_4^{-2}	$4H_2O$	
−1.05	W		$8OH^-$			−6e		WO_4^{-2}	$4H_2O$	

EMF	Elem	Anion	Compound or anion	Cation	H_2O	e =	Cation	Anion	Complex or compound	Compound or element
-1.05	Zn		$Oxin^-$			-2e			$ZnOxin^+$	
-1.04			CaH_2			-4e	$2H^+$			Ca^{+2}
-1.04	Ni		S^{-2}			-2e			$NiS(gm)$	
-1.04	Pu		Cl^-			-4e			$PuCl^{+3}$	
-1.04	Zn		$4NH_3(aq)$			-2e			$Zn(NH_3)_4^{+2}$	
-1.04	Zn		$meOxin^-$			-2e			$Zn(meOxin)^+$	
-1.03	Cd		$4CN^-$			-2e			$Cd(CN)_4^{-2}$	
-1.03	Ge		$5OH^-$			-4e	$HGeO_3^-$		$2H_2O$	
-1.02	Zn		Deta			-2e			$Zn(Deta)^{+2}$	
-1.02	Zn		$Ist(a)^-$			-2e			$Zn(Ist(a))^+$	
-1.02	Zn		$Ist(b)^-$			-2e			$Zn(Ist(b))^+$	
-1.02	Zn		Oxd^-			-2e			$ZnOxd^+$	
-1.02	Zn		$2Oxin^-$			-2e			$Zn(Oxin)_2$	
-1.01	Zn		$Himda^{-2}$			-2e			$ZnHimda$	
-1.01	Zn		$Met(a)^-$			-2e			$Zn(Met(a))^+$	
-1.01	Zn		$Met(b)^-$			-2e			$Zn(Met(b))^+$	
-1.00	In		$3OH^-$			-3e			$In(OH)_3$	
-1.00	Pa				$2H_2O$	-5e	PaO_2^+			$4H^+$
-1.00				Pa^{+3}		-e	Pa^{+4}			
-1.00			H_2Po			-2e	$2H^+$			Po
-1.00	Zn		$2meOxin^-$			-2e			$Zn(meOxin)_2$	
-1.00	Zn		$Ndpa^{-3}$			-2e			$ZnNdpa^-$	
-1.00	Zn		$2Oxd^-$			-2e			$Zn(Oxd)_2$	
-0.99	Zn		$DmHas^-$			-2e			$Zn(DmHas)^+$	
-0.99	Zn		$2DmHas^-$			-2e			$Zn(DmHas)_2$	
-0.98	Ga		$Data^{-4}$			-3e			$GaData^-$	
-0.98	Zn		Coy^-			-2e			$ZnCoy^+$	
-0.98	Zn		Has^-			-2e			$ZnHas^+$	
-0.98	Zn		$2Ist(a)^-$			-2e			$Zn(Ist(a))_2$	
-0.98	Zn		$Mcin^-$			-2e			$ZnMcin^+$	
-0.98	Zn		$Trop^-$			-2e			$ZnTrop^+$	
-0.97		CN^-	$2OH^-$			-2e		CNO^-	H_2O	

EMF	Elem	Anion	Compound or anion	Cation	H_2O	e =	Cation	Anion	Complex or compound	Compound or element
-0.97	Cd		Data^{-4}			-2e				CdData^{-2}
-0.97	H		DC14BQ$^-$			-e				DC14BQ
-0.97	Zn		Imda^{-2}			-2e				ZnImda
-0.97	Zn		2Met(a)$^-$			-2e				Zn(Met(a))$_2$
-0.97	Zn		2Ist(b)$^-$			-2e				Zn(Ist(b))$_2$
-0.97	Zn		MpHas$^-$			-2e				Zn(MpHas)$^+$
-0.97	Zn		Hox$^-$			-2e				ZnHox$^+$
-0.97	Zn		2Has$^-$			-2e				Zn(Has)$_2$
-0.97	Zn		Cin$^-$			-2e				ZnCin$^+$
-0.96		OH$^-$	Pu(OH)$_3$			-e				Pu(OH)$_4$
-0.96	Zn		2Met(b)$^-$			-2e				Zn(Met(b))$_2$
-0.96	Zn		Ptn			-2e				Zn(Ptn)$^{+2}$
-0.95	Fe		S^{-2}			-2e				FeS (a)
-0.95	Zn		Af$^-$			-2e				ZnAf$^+$
-0.95	Zn		2Mcin$^-$			-2e				Zn(Mcin)$_2$
-0.95	Zn		Ph			-2e				ZnPh^{+2}
-0.95	Zn		2Trop$^-$			-2e				Zn(Trop)$_2$
-0.94	Cr		OH$^-$			-3e				CrOH^{+2}
-0.94	Zn		Impa^{-2}			-2e				ZnImpa
-0.93	Pb		S^{-2}			-2e				PbS
-0.93		HSnO$_2^-$	3OH$^-$		H_2O	-2e		Sn(OH)$_6^{-2}$		
-0.93		SO$_3^{-2}$	2OH$^-$			-2e		SO$_4^{-2}$	H_2O	
-0.93	H		OH$^-$			-e				H_2O
-0.93	Zn		2Af$^-$			-2e				Zn(Af)$_2$
-0.93	Zn		Aspa^{-2}			-2e				ZnAspa
-0.93	Zn		2Cin$^-$			-2e				Zn(Cin)$_2$
-0.93	Zn		2Coy$^-$			-2e				Zn(Coy)$_2$
-0.93	Zn		En			-2e				ZnEn^{+2}
-0.93	Zn		2Hox$^-$			-2e				Zn(Hox)$_2$
-0.93	Zn		2MpHas$^-$			-2e				Zn(MpHas)$_2$
-0.93	Zn		2Ph			-2e				Zn(Ph)$_2^{+2}$
-0.93	Zn		Pn			-2e				ZnPn^{+2}

EMF	Elem	Anion	Compound or anion	Cation	H_2O	e =	Cation	Anion	Complex or compound	Compound or element
−0.92		Se^{-2}				−2e				Se
−0.92	Zn		2Deta			−2e			$Zn(Deta)_2^{+2}$	
−0.92	Zn		Gl^-			−2e			$ZnGl^+$	
−0.92	Zn		Dge^-			−2e			$ZnDge^+$	
−0.92	Zn		Ntp^{-3}			−2e			$ZnNtp^-$	
−0.92	H		$AlOH^-$			−e			AlOH	
−0.92	H		CH_3O^-			−e			CH_3OH	
−0.91	Be				$2H_2O$	−2e	$4H^+$	BeO_2^{-2}		
−0.91	Sn		$3OH^-$			−2e		$HSnO_2^-$	H_2O	
−0.91	Cr					−2e	Cr^{+2}			
−0.91	Zn		$AcAc^-$			−2e			$ZnAcAc^+$	
−0.91	Zn		$Alan^-$			−2e			$ZnAlan^+$	
−0.91	Zn		$2Imda^{-2}$			−2e			$Zn(Imda)_2^{-2}$	
−0.91	Zn		$Imdp^{-2}$			−2e			ZnImdp	
−0.91	Zn		$mOxin^-$			−2e			$Zn(mOxin)^+$	
−0.91	Zn		2Pn			−2e			$Zn(Pn)_2^{+2}$	
−0.90	2Tl		S^{-2}			−2e			Tl_2S	
−0.90	Zn		2En			−2e			$Zn(En)_2^{+2}$	
−0.90	Zn		Ox^{-2}			−2e			ZnOx	
−0.90	Zn		3Ph			−2e			$Zn(Ph)_3^{+2}$	
−0.89	Cd		$Edta^{-4}$			−2e			$CdEdta^{-2}$	
−0.89	2Cu		S^{-2}			−2e			Cu_2S	
−0.89		$3OH^-$	PH_3			−3e			$3H_2O$	P(white)
−0.89	Zn		$2Alan^-$			−2e			$Zn(Alan)_2$	
−0.89	Zn		$2Aspa^{-2}$			−2e			$Zn(Aspa)_2^{-2}$	
−0.89	Zn		$2Gl^-$			−2e			$Zn(Gl)_2$	
−0.89	Zn		$2Impa^{-2}$			−2e			$Zn(Impa)_2^{-2}$	
−0.89	Zn		$2mOxin^-$			−2e			$Zn(mOxin)_2$	
−0.89	Zn		OH^-			−2e			$ZnOH^+$	
−0.89	Zn		$Sald^-$			−2e			$ZnSald^+$	
−0.89	H		$Acetamide^-$			−e			Acetamide	
−0.89	H		$GlOH^-$			−e			GlOH	

EMF	Elem	Anion	Compound or anion	Cation	H_2O	e =	Cation	Anion	Complex or compound	Compound or element
−0.88	Fe		$2OH^-$			−2e			$Fe(OH)_2$	
−0.88	Ti				H_2O	−4e	TiO_2^+		$2H^+$	
−0.88	Zn		$2AcAc^-$			−2e			$Zn(AcAc)_2$	
−0.88	Zn		$2Himda^{-2}$			−2e			$Zn(Himda)_2^{-2}$	
−0.88	H		GME^-			−e			GME	
−0.87	B				$3H_2O$	−3e	$3H^+$		H_3BO_3 (c)	
−0.87	B				$3H_2O$	−3e	$3H^+$		H_3BO_3(aq)	
−0.87	Fe		$Edta^{-4}$			−2e			$FeEdta^{-2}$	
−0.87			GeH_4			−4e	$4H^+$			Ge
−0.87	Sn		S^{-2}			−2e			SnS	
−0.87	Zn		$Glgl^-$			−2e			$ZnGlgl^+$	
−0.87	Zn		$3Ist(b)^-$			−2e			$Zn(Ist(b))_3^-$	
−0.87	Zn		$3Met(b)^-$			−2e			$Zn(Met(b))_3^-$	
−0.87	Zn		$2Sald^-$			−2e			$Zn(Sald)_2$	
−0.86	Si				$2H_2O$	−4e	$4H^+$		SiO_2(quartz)	
−0.86	Zn		$2Dge^-$			−2e			$Zn(Dge)_2$	
−0.86	Zn		Mal^{-2}			−2e			ZnMal	
−0.86	Zn		$3Trop^-$			−2e			$Zn(Trop)_3^-$	
−0.86	H		ACD^-			−e			ACD	
−0.85	Zn		$2Amac^{-3}$			−2e			$Zn(Amac)_2^{-4}$	
−0.85	Zn		$2Nta^{-3}$			−2e			$Zn(Nta)_2^{-4}$	
−0.85	Zn		$SSald^{-2}$			−2e			$Zn(SSald)$	
−0.84	2B				$3H_2O$	−6e	$6H^+$			B_2O_3
−0.84	Co		$Data^{-4}$			−2e			$CoData^{-2}$	
−0.84	Zn		Ap^{-2}			−2e			ZnAp	
−0.84	Zn		$2Glgl^-$			−2e			$Zn(Glgl)_2$	
−0.84	Zn		Im			−2e			$ZnIm^{+2}$	
−0.84			$Tart^{-2}$			−2e			ZnTart	
−0.83	Cr		F^-			−3e			CrF^{+2}	
−0.83	H_2		$2OH^-$			−2e			$2H_2O$	
−0.83	Ni		S^{-2}			−2e			NiS (a)	
−0.83	Zn		$2Im$			−2e			$Zn(Im)_2^{+2}$	

EMF	Elem	Anion	Compound or anion	Cation	H_2O	e =	Cation	Anion	Complex or compound	Compound or element
-0.83	Zn		3Im			-2e			$Zn(Im)_3^{+2}$	
-0.83	Zn		NH_3			-2e			$Zn(NH_3)^{+2}$	
-0.83	Zn		$2NH_3$			-2e			$Zn(NH_3)_2^{+2}$	
-0.83	Zn		$3NH_3$			-2e			$Zn(NH_3)_3^{+2}$	
-0.83	Zn		N_2H_4			-2e			$Zn(N_2H_4)^{+2}$	
-0.83	Zn		SO_4^{-2}			-2e			$ZnSO_4$	
-0.83	Zn		$S_2O_3^{-2}$			-2e			ZnS_2O_3	
-0.83	H		TC12BQ$^-$			-e			TC12BQ	
-0.82	Zn		Gly$^-$			-2e			$ZnGly^+$	
-0.82	Zn		4Im			-2e			$Zn(Im)_4^{+2}$	
-0.82	Zn		Lac$^-$			-2e			$ZnLac^+$	
-0.82	Zn		$4NH_3$			-2e			$Zn(NH_3)_4^{+2}$	
-0.82	H		TB12BQ$^-$			-e			TB12BQ	
-0.81	Cd		$2OH^-$			-2e			$Cd(OH)_2$	
-0.81	Cr		$2F^-$			-3e			CrF_2^+	
-0.81	2Ta				$5H_2O$	-10e	$10H^+$		Ta_2O_5	
-0.81	Zn		Ac$^-$			-2e			$ZnAc^+$	
-0.81	Zn		CNS$^-$			-2e			$ZnCNS^+$	
-0.81	Zn		3En			-2e			$Zn(En)_3^{+2}$	
-0.81	Zn		Glac$^-$			-2e			$ZnGlac^+$	
-0.81	Zn		Glu$^-$			-2e			$ZnGlu^+$	
-0.81	Zn		$2N_2H_4$			-2e			$Zn(N_2H_4)_2^{+2}$	
-0.81	Zn		3Pn			-2e			$Zn(Pn)_3^{+2}$	
-0.81	Zn		Suc^{-2}			-2e			ZnSuc	
-0.81	H		CHO$^-$			-e			HCHO	
-0.80			BiH_3			-3e	$3H^+$			Bi
-0.80	Ni		Edta^{-4}			-2e			NiEdta^{-2}	
-0.80	Zn	F$^-$				-2e			ZnF^+	
-0.80	Zn		$3N_2H_4$			-2e			$Zn(N_2H_4)_3^{+2}$	
-0.80	H		NN'Gu$^-$			-e			NN'Gu	
-0.80	H		Gu$^-$			-e			Gu	
-0.80	H		MtOH$^-$			-e			MtOH	

EMF	Elem	Anion	Compound or anion	Cation	H_2O	e =	Cation	Anion	Complex or compound	Compound or element
−0.80	H		$PgOH^-$			−e			PgOH	
−0.79	4B				$7H_2O$	−12e	$14H^+$	$B_4O_7^{-2}$		
−0.79	Cr		$3F^-$			−3e			CrF_3	
−0.79			$Ti(OH)_3$			−e	H^+		TiO_2	H_2O
−0.79	Zn		Bu^-			−2e			$ZnBu^+$	
−0.79	Zn		Pr^-			−2e			$ZnPr^+$	
−0.79	H		$NNGu^-$			−e			NNGu	
−0.79	H		NGu^-			−e			NGu	
−0.78	Cd		Hed^{-3}			−2e			$CdHed^-$	
−0.78	Cr		CNS^-			−3e			$CrCNS^{+2}$	
−0.78	Fe		Hed^{-3}			−2e			$FeHed^-$	
−0.78	Si				$3H_2O$	−4e	$4H^+$			H_2SiO_3
−0.78	Zn		$3Cl^-$			−2e			$ZnCl_3^-$	
−0.78	Zn		$4N_2H_4$			−2e			$Zn(N_2H_4)_4^{+2}$	
−0.78	H		$12BQ^-$			−e			12BQ	
−0.77	H		Bzd^-			−e			Bzd	
−0.76	Cd		Tate			−2e			$Cd(Tate)^{+2}$	
−0.76	Co		$Edta^{-4}$			−2e			$CoEdta^{-2}$	
−0.76	Cr		$2CNS^-$			−3e			$Cr(CNS)_2^+$	
−0.76	Fe		CO_3^{-2}			−2e			$FeCO_3$	
−0.76	Zn					−2e	Zn^{+2}			
−0.76			Zn(Hg)			−2e	Zn^{+2}			Hg
−0.76	Zn		Nac^-			−2e			$ZnNac^+$	
−0.76	H		Dgd^-			−e			Dgd	
−0.75	Ni		Hed^{-3}			−2e			$NiHed^-$	
−0.75	Tl		I^-			−e			TlI	
−0.75	Zn		Br^-			−2e			$ZnBr^+$	
−0.75	Zn		Cl^-			−2e			$ZnCl^+$	
−0.75	Zn		$2Cl^-$			−2e			$ZnCl_2$	
−0.75	H		$Sucrose^-$			−e			Sucrose	
−0.74	Cd		CO_3^{-2}			−2e			$CdCO_3$	
−0.74	Cr					−3e	Cr^{+3}			

EMF	Elem	Anion	Compound or anion	Cation	H_2O	e =	Cation	Anion	Complex or compound	Compound or element
-0.74	Ga		OH^-			-3e			$GaOH^{+2}$	
-0.74			H_2Te (aq)			-2e	$2H^+$			Te
-0.73	Co		$2OH^-$			-2e			$Co(OH)_2$	
-0.73	Nb				H_2O	-2e	$2H^+$			NbO
-0.73	H		Acm^-			-e			Acm	
-0.73	H		$Acox^-$			-e			Acox	
-0.73	H		$CF_3CH_2O^-$			-e			CF_3CH_2OH	
-0.72	Cd		Teta			-2e			$Cd(Teta)^{+2}$	
-0.72		S^{-2}	$2FeS(a)$			-2e			Fe_2S_3	
-0.72	Ga		$AcAc^-$			-3e			$Ga(AcAc)^{+2}$	
-0.72			H_2Te (g)			-2e	$2H^+$			Te
-0.72	Ni		$2OH^-$			-2e			$Ni(OH)_2$	
-0.72	Zn		I^-			-2e			ZnI^+	
-0.72	H		$25B14B^-$			-e			25B14B	
-0.72	H		$25C14B^-$			-e			25C14B	
-0.72	H		$26C14B^-$			-e			26C14B	
-0.72	H		$P4a^-$			-e			P4a	
-0.72	H		$Glucose^-$			-e			Glucose	
-0.72	H		$CCl_3CH_2O^-$			-e			CCl_3CH_2OH	
-0.71	Co		Hed^{-3}			-2e			$CoHed^-$	
-0.71	Pb		$Data^{-4}$			-2e			$PbData^{-2}$	
-0.71	H		$23C14B^-$			-e			23C14B	
-0.71	H		$B14BQ^-$			-e			B14BQ	
-0.71	H		$C14BQ^-$			-e			C14BQ	
-0.70	Cu	S^{-2}				-2e				CuS
-0.70	Fe		Nta^{-3}			-2e			$FeNta^-$	
-0.70	Fe		Tate			-2e			$Fe(Tate)^{+2}$	
-0.69			BaH_2			-2e	$2H^+$			Ba
-0.69	Ga		$2AcAc^-$			-3e			$Ga(AcAc)_2^+$	
-0.69	Hg		S^{-2}			-2e			HgS	
-0.69	Ni		Tate			-2e			$Ni(Tate)^{+2}$	
-0.69	H		Bzm^-			-e			Bzm	

EMF	Elem	Anion	Compound or anion	Cation	H_2O	e =	Cation	Anion	Complex or compound	Compound or element
-0.69			2OHQ			-e	H^+	2OHQ⁻		
-0.68	As		4OH⁻			-3e		AsO_2^-	$2H_2O$	
-0.68		AsO_2^-	4OH⁻			-2e		AsO_4^{-3}	$2H_2O$	
-0.68	Cd		Nta^{-3}			-2e			CdNta⁻	
-0.68	Cd		Oxin⁻			-2e			$CdOxin^+$	
-0.68				Ga^{+2}		-e	Ga^{+3}			
-0.68	H		Glt⁻			-e			Glt	
-0.67	Au	2CN⁻				-e		$Au(CN)_2^-$		
-0.67	Fe		Teta			-2e			$Fe(Teta)^{+2}$	
-0.67	Pb		$Edta^{-4}$			-2e			$PbEdta^{-2}$	
-0.67	H		Azt⁻			-e			Azt	
-0.67	H		Chl⁻			-e			Chl	
-0.67	H		Pld⁻			-e			Pld	
-0.66	2Ag		S^{-2}			-2e			Ag_2S	
-0.66	Co		Tate			-2e			$Co(Tate)^{+2}$	
-0.66	Ni		Teta			-2e			$Ni(Teta)^{+2}$	
-0.66	Sb		4OH⁻			-3e		SbO_2^-	$2H_2O$	
-0.66	Tl		Br⁻			-e			TlBr	
-0.66	H		2B5M14⁻			-e			2B5M14	
-0.66			4OHQ			-e	H^+	4OHQ⁻		
-0.66	H		Mln⁻			-e			Mln	
-0.66	H		12N46D⁻			-e			12N46D	
-0.66	H		Pip⁻			-e			Pip	
-0.65	Cd		Deta			-2e			$Cd(Deta)^{+2}$	
-0.65	Ga		3AcAc⁻			-3e			$Ga(AcAc)_3$	
-0.65	H		2C5M14⁻			-e			2C5M14	
-0.65	H		Dea⁻			-e			Dea	
-0.65	H		18Oda⁻			-e			18Oda	
-0.64	Co		CO_3^{-2}			-2e			$CoCO_3$	
-0.64	Cu	CN⁻				-e				CuCN
-0.64	2Nb				$5H_2O$	-10e	$10H^+$		Nb_2O_5	
-0.64			5Avl			-e	H^+	5Avl⁻		

EMF	Elem	Anion	Compound or anion	Cation	H_2O	e =	Cation	Anion	Complex or compound	Compound or element
−0.64	H		14Bda⁻			−e			14Bda	
−0.64	H		Dma⁻			−e			Dma	
−0.64	H		2M14BQ⁻			−e			2M14BQ	
−0.64	H		Tea⁻			−e			Tea	
−0.63	Cd		2Oxin⁻			−2e			$Cd(Oxin)_2$	
−0.63	Ga		F⁻			−3e			GaF^{+2}	
−0.63			NbO		H_2O	−2e	$2H^+$			NbO_2
−0.63	H		Bla⁻			−e			Bla	
−0.63	H		Cha⁻			−e			Cha	
−0.63	H		Dcsa⁻			−e			Dcsa	
−0.63	H		Dda⁻			−e			Dda	
−0.63	H		EAcAc⁻			−e			EAcAc	
−0.63	H		Ea⁻			−e			Ea	
−0.63	H		Hda⁻			−e			Hda	
−0.63	H		Ma⁻			−e			Ma	
−0.63	H		12N4S⁻			−e			12N4S	
−0.63	H		Oa⁻			−e			Oa	
−0.63	H		13Pda⁻			−e			13Pda	
−0.63	H		iPa⁻			−e			iPa	
−0.63	H		Uda⁻			−e			Uda	
−0.62	Cd		Ndap⁻³			−2e			CdNdap⁻	
−0.62	2Cd	(FeCy)⁻⁴				−4e				$Cd_2(FeCy)$
−0.62	Fe		Deta			−2e			$Fe(Deta)^{+2}$	
−0.62			mOxin⁻	UO_2^+		−e			$UO_2(mOxin)^+$	
−0.62			meOxin⁻	UO_2^+		−e			$UO_2(meOxin)^+$	
−0.62			Oxin⁻	UO_2^+		−e			$UO_2(Oxin)^+$	
−0.62	H		tBa⁻			−e			tBa	
−0.62	H		$C_2H_5S^-$			−e			C_2H_5SH	
−0.62	H		NMP1d⁻			−e			NMP1d	
−0.62			2Mea			−e	H^+	2Mea⁻		
−0.62	H		nPa⁻			−e			nPa	
−0.62	H		TGa⁻			−e			TGa	

EMF	Elem	Anion	Compound or anion	Cation	H_2O	e =	Cation	Anion	Complex or compound	Compound or element
-0.61			AsH_3 (g)			-3e	$3H^+$			As
-0.61	Cd		$4NH_3$ (aq)			-2e			$Cd(NH_3)_4^{+2}$	
-0.61	Cd		$Himda^{-2}$			-2e			CdHimda	
-0.61	Co		Nta^{-3}			-2e			$CoNta^-$	
-0.61	Co		Teta			-2e			$Co(Teta)^{+2}$	
-0.61	Fe		Ph			-2e			$FePh^{+2}$	
-0.61				U^{+3}		-e	U^{+4}			
-0.61			4AP			-e	H^+	$4AP^-$		
-0.61			3APA			-e	H^+	$3APA^-$		
-0.61	H		oC^-			-e				oC
-0.61	H		Cna^-			-e				Cna
-0.60	Cd		$2Amac^{-3}$			-2e			$Cd(Amac)_2^{-4}$	
-0.60	Cd		Coy^-			-2e			$CdCoy^+$	
-0.60	H		mC^-			-e				mC
-0.60	H		pC^-			-e				pC
-0.60	H		$sDPGu^-$			-e				sDPGu
-0.60	H		$12Eda^-$			-e				12Eda
-0.60	H		$2MPa^-$			-e				2MPa
-0.60			2SQ			-e	H^+	$2SQ^-$		
-0.60	H		pMP^-			-e				pMP
-0.60	H		$NMPip^-$			-e				NMPip
-0.60	H		NOM^-			-e				NOM
-0.59	Cd		Ptn			-2e			$Cd(Ptn)^{+2}$	
-0.59	Co		$Oxin^-$			-2e			$CoOxin^+$	
-0.59	Fe		2Dyp			-2e			$Fe(Dyp)_2^{+2}$	
-0.59	H		IO^-			-e				HIO
-0.59	Ni		$Ndap^{-3}$			-2e			$NiNdap^-$	
-0.59	Ni		$Oxin^-$			-2e			$NiOxin^-$	
-0.59		$4OH^-$	ReO_2			-3e		ReO_4^-	$2H_2O$	
-0.59			$MeOxin^-$	UO_2^+		-e			$UO_2(MeOxin)^+$	
-0.59			2Apa			-e	H^+	$2Apa^-$		
-0.59	H		$12CHDA^-$			-e				12CHDA

EMF	Elem	Anion	Compound or anion	Cation	H_2O	e =	Cation	Anion	Complex or compound	Compound or element
-0.59	H		$25M14B^-$			-e			25M14B	
-0.59	H		$23M14B^-$			-e			23M14B	
-0.59	H		$25DMBQ^-$			-e			25DMBQ	
-0.59	H		pFP^-			-e			pFP	
-0.59	H		$OH14BQ^-$			-e			OH14BQ	
-0.59	H		$CX14BQ^-$			-e			CX14BQ	
-0.59	H		oMP^-			-e			oMP	
-0.59	H		$2M5114^-$			-e			2M5114	
-0.59	H		$Phenol^-$			-e			Phenol	
-0.59	H		$Peba^-$			-e			Peba	
-0.59	H		$oPhP^-$			-e			oPhP	
-0.59	H		$Quinol^-$			-e			Quinol	
-0.59			Oxin			-e	H^+	$Oxin^-$		
-0.58	Co		$Ndap^{-3}$			-2e			$CoNdap^-$	
-0.58	2Cr				$3H_2O$	-6e	$6H^+$			Cr_2O_3
-0.58	Fe		Ox^{-2}			-2e			FeOx	
-0.58	Ni		Nta^{-3}			-2e			$NiNta^-$	
-0.58	Pb		$2OH^-$			-2e			PbO (r)	H_2O
-0.58	Re		$4OH^-$			-4e			ReO_2	$2H_2O$
-0.58	Re		$8OH^-$			-7e		ReO_4^-		$4H_2O$
-0.58			2AAa			-e	H^+	$2AAa^-$		
-0.58			3AP			-e	H^+	$3AP^-$		
-0.58	H		$OMiU^-$			-e			OMiU	
-0.58	H		$SMiT^-$			-e			SMiT	
-0.58	H		$1-Np^-$			-e			1-Np	
-0.58	H		Pea^-			-e			Pea	
-0.58	H		$Pipz^-$			-e			Pipz	
-0.58	H		Tma^-			-e			Tma	
-0.57	H		AsO_2^-			-e			$HAsO_2$	
-0.57	Cd		$Ndpa^{-3}$			-2e			$CdNdpa^-$	
-0.57	Cd		$2Nta^{-3}$			-2e			$Cd(Nta)_2^{-4}$	
-0.57	Co		Oxd^-			-2e			$CoOxd^+$	

EMF	Elem	Anion	Compound or anion	Cation	H_2O	e =	Cation	Anion	Complex or compound	Compound or element
-0.57	Fe		Dge⁻			-2e			FeDge⁺	
-0.57	Fe		En			-2e			FeEn⁺²	
-0.57	Ni		Deta			-2e			Ni(Deta)⁺²	
-0.57		$S_2O_3^{-2}$	6OH⁻			-4e		$2SO_3^{-2}$	$3H_2O$	
-0.57	Te		6OH⁻			-4e		TeO_3^{-2}	$3H_2O$	
-0.57	H		Alma⁻			-e			Alma	
-0.57			2AP			-e	H⁺	2AP⁻		
-0.57	H		13D2P⁻			-e			13D2P	
-0.57	H		mMP⁻			-e			mMP	
-0.57	H		2MTHP⁻			-e			2MTHP	
-0.57	H		2Np⁻			-e			2Np	
-0.57	H		mPhP⁻			-e			mPhP	
-0.57	H		pPhP⁻			-e			pPhP	
-0.57	H		Sccm⁻			-e			Sccm	
-0.56	Cd		CN⁻			-2e			CdCN⁺	
-0.56	Cd		2Deta			-2e			Cd(Deta)₂⁺²	
-0.56	Cd		En			-2e			CdEn⁺²₂	
-0.56	Cd		Imda⁻²			-2e			CdImda	
-0.56	Fe		2Deta			-2e			Fe(Deta)₂⁺²	
-0.56	Fe		Dyp			-2e			FeDyp⁺²	
-0.56	Fe		OH⁻			-2e			FeOH⁺	
-0.56		OH⁻	Fe(OH)₂			-e			Fe(OH)₃	
-0.56	Fe		Sald⁻			-2e			FeSald⁺	
-0.56		O_2^-				-e				O_2
-0.56	H		Ahoa⁻			-e			Ahoa	
-0.56	H		Bzs⁻			-e			Bzs	
-0.56	H		Ctch⁻			-e			Ctch	
-0.56	H		pCP⁻			-e			pCP	
-0.56	H		Eoa⁻			-e			Eoa	
-0.56	H		Hem⁻			-e			Hem	
-0.56	H		Mesa⁻			-e			Mesa	

EMF	Elem	Anion	Compound or anion	Cation	H_2O	e =	Cation	Anion	Complex or compound	Compound or element
-0.56	H		Mxea⁻			-e			Mxea	
-0.56	H		mMSP⁻			-e			mMSP	
-0.56	H		pMSP⁻			-e			pMSP	
-0.56	H		Rsl⁻			-e			Rsl	
-0.55	Cd		2CN⁻			-2e			$Cd(CN)_2$	
-0.55	Cd		2Himda⁻²			-2e			$Cd(Himda)_2^{-2}$	
0.55	Ga	Cl⁻				-3e			$GaCl^{+2}$	
-0.55	Co		2Oxin⁻			-2e			$Co(Oxin)_2$	
-0.55	Tl	N₃⁻				-e				TlN_3
-0.55			Coy⁻	UO_2^+		-e			$UO_2(Coy)^+$	
-0.55	H		Bzma⁻			-e			Bzma	
-0.55	H		pBP⁻			-e			pBP	
-0.55	H		pIP⁻			-e			pIP	
-0.55	H		mFP⁻			-e			mFP	
-0.55	H		12NQ⁻			-e			12NQ	
-0.54	H		$H_2BO_3^-$			-e			H_3BO_3	
-0.54	Cd		3CN⁻			-2e			$Cd(CN)_3^-$	
-0.54	Cd		2Coy⁻			-2e			$Cd(Coy)_2$	
-0.54	Cd		Dge⁻			-2e			$Cd(Dge)^+$	
-0.54	Co		2Oxd⁻			-2e			$Co(Oxd)_2$	
-0.54	Fe		2En			-2e			$Fe(En)_2^{+2}$	
-0.54	Fe		2Sald⁻			-2e			$Fe(Sald)_2$	
-0.54	H		CN⁻			-e			HCN	
-0.54	In		OH⁻			-3e			$InOH^{+2}$	
-0.54	Ni		2Oxin⁻			-2e			$Ni(Oxin)_2$	
-0.54	Pb		3OH⁻			-2e		$HPbO_2^-$	H_2O	
-0.54	H		maP⁻			-e			maP	
-0.54			AESA			-e	H^+	AESA⁻		
-0.54	H		4APy⁻			-e			4APy	
-0.54	H		Chxm⁻			-e			Chxm	
-0.54	H		Ecea⁻			-e			Ecea	
-0.54	H		mIP⁻			-e			mIP	

EMF	Elem	Anion	Compound or anion	Cation	H_2O	e =	Cation	Anion	Complex or compound	Compound or element
-0.53	Cd		$Aspa^{-2}$			-2e				$CdAspa$
-0.53	Cd		$2En$			-2e				$Cd(En)_2^{+2}$
-0.53	Cd		$Impa^{-2}$			-2e				$CdImpa$
-0.53	Cd		Pn			-2e				$CdPn^{+2}$
-0.53	Cd		$2Pn$			-2e				$Cd(Pn)_2^{+2}$
-0.53	Fe		$2Dge^-$			-2e				$Fe(Dge)_2$
-0.53	Ga					-3e	Ga^{+3}			
-0.53	Ni		$Himda^{-2}$			-2e				$NiHimda$
-0.53	Ni		Ptn			-2e				$NiPtn^{+2}$
-0.53			$2meOxin^-$	UO_2^+		-e				$UO_2(meOxin)_2$
-0.53	H		$Bhoa^-$			-e				$Bhoa$
-0.53	H		mBP^-			-e				mBP
-0.53	H		mCP^-			-e				mCP
-0.53	H		$mFmP^-$			-e				$mFmP$
-0.53			$6OHQ$			-e	H^+	$6OHQ^-$		
-0.53	H		$14NQ2S^-$			-e				$14NQ2S$
-0.53	H		$14NQ3S^-$			-e				$14NQ3S$
-0.53	H		$235MBQ^-$			-e				$235MBQ$
-0.52	Cd		Cit^{-3}			-2e				$CdCit^-$
-0.52	Cd		$2Imda^{-2}$			-2e				$Cd(Imda)_2^{-2}$
-0.52	Cd		$S_2O_3^{-2}$			-2e				CdS_2O_3
-0.52	Co		$Deta$			-2e				$Co(Deta)^{+2}$
-0.52	Co		$Himda^{-2}$			-2e				$CoHimda$
-0.52	Co		$Ist(a)^-$			-2e				$Co(Ist(a))^+$
-0.52	Co		$Met(a)^-$			-2e				$Co(Met(a))^+$
-0.52	Fe		$3En$			-2e				$Fe(En)_3^{+2}$
-0.52	Fe		Mal^{-2}			-2e				$FeMal$
-0.52	Ni		$Ndpa^{-3}$			-2e				$NiNdpa^-$
-0.52			$2Oxin^-$	UO_2^+		-e				$UO_2(Oxin)_2$
-0.52	H		oFP^-			-e				oFP
-0.52			$4SQ$			-e	H^+	$4SQ^-$		
-0.52	H		$PhBA^-$			-e				$PhBA$

EMF	Elem	Anion	Compound or anion	Cation	H_2O	e =	Cation	Anion	Complex or compound	Compound or element
-0.52			7OHQ			-e	H^+	$7OHQ^-$		
-0.51	Cd		$AcAc^-$			-2e			$CdAcAc^+$	
-0.51	Cd		$4CN^-$			-2e			$Cd(CN)_4^{-2}$	
-0.51	Co		$Ist(b)^-$			-2e			$Co(Ist(b))^+$	
-0.51	Co		$Met(b)^-$			-2e			$Co(Met(b))^+$	
-0.51	Co		$Ndpa^{-3}$			-2e			$CoNdpa^-$	
-0.51	Fe		SO_4^{-2}			-2e			$FeSO_4$	
-0.51	Ga	$2Cl^-$				-3e			$GaCl_2^+$	
-0.51	H		BrO^-			-e			$HBrO$	
-0.51	Nd		$2AcAc^-$			-3e			$Nd(AcAc)_2^+$	
-0.51	P				$2H_2O$	-e	H^+		H_3PO_2	
-0.51	Pb		CO_3^{-2}			-2e			$PbCO_3$	
-0.51			SbH_3 (g)			-3e	$3H^+$			Sb
-0.51			$2mOxin^-$	UO_2^+		-e			$UO_2(mOxin)_2$	
-0.51			Oxd^-	UO_2^+		-e			$UO_2(Oxd)^+$	
-0.51	H		$mCNP^-$			-e			$mCNP$	
-0.51	H		$26M14B^-$			-e			$26M14B$	
-0.51	H		$Mphl^-$			-e			$Mphl$	
-0.51			5OHQ			-e	H^+	$5OHQ^-$		
-0.50	Cd		$2Dge^-$			-2e			$Cd(Dge)_2$	
-0.50	Cd		$Imdp^{-2}$			-2e			$CdImdp$	
-0.50	Cd		Mal^{-2}			-2e			$CdMal$	
-0.50	Cd		Ntp^{-3}			-2e			$CdNtp^-$	
-0.50	Cd		Ox^{-2}			-2e			$CdOx$	
-0.50	Fe		$S_2O_3^{-2}$			-2e			FeS_2O_3	
-0.50			H_3PO_2(aq)		H_2O	-2e	$2H^+$		H_3PO_3(aq)	
-0.50	In		$AcAc^-$			-3e			$InAcAc^{+2}$	
-0.50	Ni		$Ist(a)^-$			-2e			$Ni(Ist(a))^+$	
-0.50	Ni		$Ist(b)^-$			-2e			$Ni(Ist(b))^+$	
-0.50	Ni		$Mcin^-$			-2e			$Ni(Mcin)^+$	
-0.50	Ni		$Met(a)^-$			-2e			$Ni(Met(a))^+$	
-0.50	Ni		$Met(b)^-$			-2e			$Ni(Met(b))^+$	

EMF	Elem	Anion	Compound or anion	Cation	H_2O	e =	Cation	Anion	Complex or compound	Compound or element
−0.50	Ni		Oxd^-			−2e			$NiOxd^+$	
−0.50				Pa^{+3}	$2H_2O$	−e	$4H^+$			PaO_2
−0.50			$2MeOxin^-$	UO_2^+		−e			$UO_2(MeOxin)_2$	
−0.50	H		oCP^-			−e				oCP
−0.50	H		oBP^-			−e				oBP
−0.50	H		$oFmP^-$			−e				$oFmP$
−0.50	H		oIP^-			−e				oIP
−0.50	H		$pMCP^-$			−e				$pMCP$
−0.50	H		$mMsP^-$			−e				$mMsP$
−0.50	H		$NOET^-$			−e				$NOET$
−0.50	H		$mNOP^-$			−e				$mNOP$
−0.49	Cd		$2Aspa^{-2}$			−2e			$Cd(Aspa)_2^{-2}$	
−0.49	Cd		$2Impa^{-2}$			−2e			$Cd(Impa)_2^{-2}$	
−0.49	Co		$Imda^{-2}$			−2e			$CoImda$	
−0.49	Co		$Trop^-$			−2e			$CoTrop^+$	
−0.49	2Ga				$3H_2O$	−6e	$6H^+$			Ga_2O_3
−0.49	Ni		Cin^-			−2e			$NiCin^+$	
−0.49				In^{+2}		−e	In^{+3}			
−0.49	Ni		$2Deta$			−2e			$Ni(Deta)_2^{+2}$	
−0.49	Ni		$Imda^{-2}$			−2e			$NiImda$	
−0.49	Ni		$2Mcin^-$			−2e			$Ni(Mcin)_2$	
−0.49	Ni		$2Oxd^-$			−2e			$Ni(Oxd)_2$	
−0.49	Po		$6OH^-$			−4e		PoO_3^{-2}	$3H_2O$	
−0.49	H		$AcAc^-$			−e				$AcAc$
−0.49	H		$NAcGu^-$			−e				$NAcGu$
−0.49	H		$Bzac^-$			−e				$Bzac$
−0.49	H		$2Mea^-$			−e				$2Mea$
−0.48	Cd		$2AcAc^-$			−2e			$Cd(AcAc)_2$	
−0.48	Cd		Im			−2e			$Cd(Im)^{+2}$	
−0.48	Cd		CH_3NH_2			−2e			$Cd(CH_3NH_2)^{+2}$	
−0.48	Cd		NH_3			−2e			$Cd(NH_3)^{+2}$	

EMF	Elem	Anion	Compound or anion	Cation	H_2O	e =	Cation	Anion	Complex or compound	Compound or element
-0.48	Co		Coy^-			-2e			$Co(Coy)^+$	
-0.48	Co		$2Ist(a)^-$			-2e			$Co(Ist(a))_2$	
-0.48	Co		Ptn			-2e			$Co(Ptn)^{+2}$	
-0.48	Fe		NH_3			-2e			$Fe(NH_3)^{+2}$	
-0.48	In		$2AcAc^-$			-2e			$In(AcAc)_2^+$	
-0.48	Ni		$6NH_3$ (aq)			-2e			$Ni(NH_3)_6^{+2}$	
-0.48	Ni		$DmHas^-$			-2e			$Ni(DmHas)^+$	
-0.48	Ni		En			-2e			$Ni(En)^{+2}$	
-0.48	Ni		Hox^-			-2e			$Ni(Hox)^+$	
-0.48	Ni		$Trop^-$			-2e			$Ni(Trop)^+$	
-0.48	Pb		Nta^{-3}			-2e			$PbNta^-$	
-0.48			Has^-	UO_2^+		-e			$UO_2(Has)^+$	
-0.48			$Mcin^-$	UO_2^+		-e			$UO_2(Mcin)^+$	
-0.48	H		paP^-			-e			paP	
-0.48	H		$Azrd^-$			-e			Azrd	
-0.48			3OHQ			-e	H^+	$3OHQ^-$		
-0.47	Cd		OH^-			-2e			$CdOH^+$	
-0.47	Cd		I^-			-2e			CdI^+	
-0.47	Cd		$3I^-$			-2e			CdI_3^-	
-0.47	Cd		SO_4^{-2}			-2e			$CdSO_4$	
-0.47	Cd		$2S_2O_3^{-2}$			-2e			$Cd(S_2O_3)_2^{-2}$	
-0.47	Co		$2Ist(b)^-$			-2e			$Co(Ist(b))_2$	
-0.47	Co		$2Met(a)^-$			-2e			$Co(Met(a))_2$	
-0.47	Ga	$3Cl^-$				-3e				$GaCl_3$
-0.47	Ni		$Impa^{-2}$			-2e			NiImpa	
-0.47	Ni		Pn			-2e			$Ni(Pn)^{+2}$	
-0.47			$DmHas^-$	UO_2^+		-e			$UO_2(DmHas)^+$	
-0.47	H		$Cbma^-$			-e			Cbma	
-0.47	H		$pCNP^-$			-e			pCNP	
-0.47	H		$14NQ^-$			-e			14NQ	
-0.46	Bi		$6OH^-$			-6e			Bi_2O_3	$3H_2O$

EMF	Elem	Anion	Compound or anion	Cation	H_2O	e =	Cation	Anion	Complex or compound	Compound or element
-0.46	Cd		$2CH_3NH_2$			-2e			$Cd(CH_3NH_2)_2^{+2}$	
-0.46	Cd		3En			-2e			$Cd(En)_3^{+2}$	
-0.46	Cd		2Im			-2e			$Cd(Im)_2^{+2}$	
-0.46	Cd		$2NH_3$			-2e			$Cd(NH_3)_2^{+2}$	
-0.46	Cd		3Pn			-2e			$Cd(Pn)_3^{+2}$	
-0.46	Co		2Deta			-2e			$Co(Deta)_2^{+2}$	
-0.46	Co		$Impa^{-2}$			-2e			CoImpa	
-0.46	Co		$2Met(b)^-$			-2e			$Co(Met(b))_2$	
-0.46	Fe		$2NH_3$			-2e			$Fe(NH_3)_2^{+2}$	
-0.46	Fe		$3Ox^{-2}$			-2e			$Fe(Ox)_3^{-4}$	
-0.46	Ni		$Aspa^{-2}$			-2e			NiAspa	
-0.46	Ni		$2Cin^-$			-2e			$Ni(Cin)_2$	
-0.46	Ni		Coy^-			-2e			$Ni(Coy)^+$	
-0.46	Ni		$2DmHas^-$			-2e			$Ni(DmHas)_2$	
-0.46	Ni		$2Hox^-$			-2e			$Ni(Hox)_2$	
-0.46			Cin^-	UO_2^+		-e			$UO_2(Cin)^+$	
-0.46	H		$CNea^-$			-e			CNea	
-0.46	H		$25E14B^-$			-e			25E14B	
-0.46	H		$pMsP^-$			-e			pMsP	
-0.46	H		$Trma^-$			-e			Trma	
-0.45	Cd		Ac^-			-2e			$CdAc^+$	
-0.45	Cd		Br^-			-2e			$CdBr^+$	
-0.45	Cd		CSN_2H_4			-2e			$Cd(CSN_2H_4)^{+2}$	
-0.45	Cd		$2I^-$			-2e			CdI_2	
-0.45	Cd		3Im			-2e			$Cd(Im)_3^{+2}$	
-0.45	Cd		$2Ox^{-2}$			-2e			$Cd(Ox)_2^{-2}$	
-0.45	Co		Af^-			-2e			$CoAf^+$	
-0.45	Co		$Aspa^{-2}$			-2e			CoAspa	
-0.45	Co		En			-2e			$Co(En)^{+2}$	
-0.45	Co		$2Trop^-$			-2e			$Co(Trop)_2$	
-0.45	Ga	$4Cl^-$				-3e			$GaCl_4^-$	

EMF	Elem	Anion	Compound or anion	Cation	H_2O	e =	Cation	Anion	Complex or compound	Compound or element
-0.45	Ni		CO_3^{-2}			-2e			$NiCO_3$	
-0.45	Ni		Af^-			-2e			$NiAf^+$	
-0.45	Ni		$2Ist(a)^-$			-2e			$Ni(Ist(a))_2$	
-0.45	Ni		$2Met(a)^-$			-2e			$Ni(Met(a))_2$	
-0.45	Ni		$2Met(b)^-$			-2e			$Ni(Met(b))_2$	
-0.45	P				$3H_2O$	-3e	$3H^+$			H_3PO_3
-0.45			S^{-2}			-2e				S
-0.45	H		$HTeO_4^-$			-e			H_2TeO_4	
-0.45	Tl		$S_2O_3^{-2}$			-e			$TlS_2O_3^-$	
-0.45			Hox^-	UO_2^+		-e			$UO_2(Hox)^+$	
-0.45			$MpHas^-$	UO_2^+		-e			$UO_2(MpHas)^+$	
-0.45	H		$pFmP^-$			-e			$pFmP$	
-0.45	H		$Mcma^-$			-e			$Mcma$	
-0.45	H		$MSGly^-$			-e			$MSGly$	
-0.44	Cd		CNS^-			-2e			$CdCNS^+$	
-0.44	Cd		$4I^-$			-2e			CdI_4^{-2}	
-0.44	Cd		$3NH_3$			-2e			$Cd(NH_3)_3^{+2}$	
-0.44	Co		$AcAc^-$			-2e			$Co(AcAc)^+$	
-0.44	Co		Dge^-			-2e			$Co(Dge)^+$	
-0.44	Co		$2Imda^{-2}$			-2e			$Co(Imda)_2^{-2}$	
-0.44	Fe					-2e	Fe^{+2}			
-0.44				In^+		-2e	In^{+3}			
-0.44	Ni		Dge^-			-2e			$Ni(Dge)^+$	
-0.44	Ni		$2En$			-2e			$Ni(En)_2^{+2}$	
-0.44	Ni		$2Imda^{-2}$			-2e			$Ni(Imda)_2^{-2}$	
-0.44	Ni		$2Ist(b)^-$			-2e			$Ni(Ist(b))_2$	
-0.44	Ni		$2Pn$			-2e			$Ni(Pn)_2^{+2}$	
-0.44	Ni		$Tmen$			-2e			$Ni(Tmen)^{+2}$	
-0.44	Pb		Oxd^-			-2e			$PbOxd^+$	
-0.44	Pb		$Oxin^-$			-2e			$PbOxin^+$	
-0.44	Tl		$P_2O_7^{-4}$			-e			$TlP_2O_7^{-3}$	
-0.44	Zn				H_2O	-2e	$2H^+$			ZnO

EMF	Elem	Anion	Compound or anion	Cation	H_2O	e =	Cation	Anion	Complex or compound	Compound or element
-0.44	H		$DAcAc^-$			-e				DAcAc
-0.44	H		$25H14B^-$			-e				25H14B
-0.44	H		$OHClBQ^-$			-e				OHClBQ
-0.43	Cd		$2Ac^-$			-2e				$Cd(Ac)_2$
-0.43	Cd		$3Br^-$			-2e				$CdBr_3^-$
-0.43	Cd .		$3CH_3NH_2$			-2e				$Cd(CH_3NH_2)_3^{+2}$
-0.43	Cd		$4CNS^-$			-2e				$Cd(CNS)_4^{-2}$
-0.43	Cd		$2CSN_2H_4$			-2e				$Cd(CSN_2H_4)_2^{+2}$
-0.43	Cd		$4Im$			-2e				$Cd(Im)_4^{+2}$
-0.43	Cd		$4NH_3$			-2e				$Cd(NH_3)_4^{+2}$
-0.43	Co		$2Coy^-$			-2e				$Co(Coy)_2$
-0.43	Co		Gl^-			-2e				$CoGl^+$
-0.43	Co		$Imdp^{-2}$			-2e				$CoImdp$
-0.43	Cu		$2CN^-$			-e				$Cu(CN)_2^-$
-0.43				Eu^{+2}		-e	Eu^{+3}			
-0.43	H		ClO^-			-e				HClO
-0.43	6Hg		$2(CoCy)^{-3}$			-6e				$(Hg_2)_3(CoCy)_2$
-0.43	Ni		$AcAc^-$			-2e				$Ni(AcAc)^+$
-0.43	Ni		$Alan^-$			-2e				$Ni(Alan)^+$
-0.43	Ni		$Imdp^{-2}$			-2e				$NiImdp$
-0.43	Ni		$2Trop^-$			-2e				$Ni(Trop)_2$
-0.43	Sn		OH^-			-2e				$SnOH^+$
-0.43	H		$oNOP^-$			-e				oNOP
-0.42	Cd		$2Br^-$			-2e				$CdBr_2$
-0.42	Cd		$4CH_3NH_2$			-2e				$Cd(CH_3NH_2)_4^{+2}$
-0.42	Cd		$2Cl^-$			-2e				$CdCl_2$
-0.42	Cd		$2CNS^-$			-2e				$Cd(CNS)_2$
-0.42	Cd		$3CNS^-$			-2e				$Cd(CNS)_3^-$

EMF	Elem	Anion	Compound or anion	Cation	H_2O	e =	Cation	Anion	Complex or compound	Compound or element
-0.42	Co		$2Af^-$			-2e			$Co(Af)_2$	
-0.42	Co		$Alan^-$			-2e			$Co(Alan)^+$	
-0.42	Co		$2En$			-2e			$Co(En)_2^{+2}$	
-0.42	Co		Ntp^{-3}			-2e			$CoNtp^-$	
-0.42	Co		Ox^{-2}			-2e			$CoOx$	
-0.42	Co		$Sald^-$			-2e			$Co(Sald)^+$	
-0.42	Ni		Gl^-			-2e			$Ni(Gl)^+$	
-0.42	Ni		Ntp^{-3}			-2e			$NiNtp^-$	
-0.42	Ni		$P_2O_7^{-4}$			-2e			$NiP_2O_7^{-2}$	
-0.42			$2Oxd^-$	UO_2^+		-e			$UO_2(Oxd)_2$	
-0.42	H		$NDot^-$			-e			$NDot$	
-0.42	H		Im^-			-e			Im	
-0.42	H		$pNOP^-$			-e			$pNOP$	
-0.42	H		$Ntba^-$			-e			$Ntba$	
-0.41	Cd		$4Br^-$			-2e			$CdBr_4^{-2}$	
-0.41	Cd		$3Cl^-$			-2e			$CdCl_3^-$	
-0.41	Cd		Nac^-			-2e			$Cd(Nac)^+$	
-0.41	Cd		NO_3^-			-2e			$CdNO_3^+$	
-0.41	Co		$2Aspa^{-2}$			-2e			$Co(Aspa)_2^{-2}$	
-0.41	Co		$2Himda^{-2}$			-2e			$Co(Himda)_2^{-2}$	
-0.41	Co		$2Impa^{-2}$			-2e			$Co(Impa)_2^{-2}$	
-0.41	Co		OH^-			-2e			$CoOH^+$	
-0.41				Cr^{+2}		-e	Cr^{+3}			
-0.41	H		HS^-			-e			H_2S	
-0.41	2H		S^{-2}			-2e			H_2S	
-0.41	In		Ac^-			-3e			$In(Ac)^{+2}$	
-0.41	In		F^-			-3e			InF^{+2}	
-0.41	Ni		$2Aspa^{-2}$			-2e			$Ni(Aspa)_2^{-2}$	
-0.41	Ni		$2Coy^-$			-2e			$Ni(Coy)_2$	
-0.41	Ni		Ox^{-2}			-2e			$NiOx$	
-0.41	Pb		$Himda^{-2}$			-2e			$PbHimda$	

EMF	Elem	Anion	Compound or anion	Cation	H_2O	e =	Cation	Anion	Complex or compound	Compound or element
−0.41	Pb		$Ist(a)^-$			−2e			$Pb(Ist(a))^+$	
−0.41	Pb		$Met(a)^-$			−2e			$Pb(Met(a))^+$	
−0.41	Pb		$Met(b)^-$			−2e			$Pb(Met(b))^+$	
−0.41			$2Has^-$	UO_2^+		−e			$UO_2(Has)_2$	
−0.41			$2MpHas^-$	UO_2^+		−e			$UO_2(MpHas)_2$	
−0.41	H		$2APy^-$			−e			$2APy$	
−0.41	2H		Sal^{-2}			−2e			Sal	
−0.40	Cd					−2e	Cd^{+2}			
−0.40	Co		$2AcAc^-$			−2e			$Co(AcAc)_2$	
−0.40	Co		$2Gl^-$			−2e			$Co(Gl)_2$	
−0.40	Co		$2Nta^{-3}$			−2e			$Co(Nta)_2^{-4}$	
−0.40	2Ga				H_2O	−2e	$2H^+$			Ga_2O
−0.40			H_2Se (aq)			−2e	$2H^+$			Se
−0.40				In^+		−e	In^{+2}			
−0.40	Ni		$2Himda^{-2}$			−2e			$Ni(Himda)_2^{-2}$	
−0.40	Ni		$2Impa^{-2}$			−2e			$Ni(Impa)_2^{-2}$	
−0.40	Ni		$P_4O_{12}^{-4}$			−2e			$NiP_4O_{12}^{-2}$	
−0.40	Ni		$Sald^-$			−2e			$Ni(Sald)^+$	
−0.40	Pb		$Ist(b)^-$			−2e			$Pb(Ist(b))^+$	
−0.40	Pd	$4CN^-$				−2e			$Pd(CN)_4^{-2}$	
−0.40			$AcAc^-$	UO_2^+		−e			$UO_2(AcAc)^+$	
−0.40		Re^-				−e				Re
−0.40	H		DPK^-			−e			DPK	
−0.39	Cd		$5NH_3$			−2e			$Cd(NH_3)_5^{+2}$	
−0.39	Co		$2Alan^-$			−2e			$Co(Alan)_2$	
−0.39	Co		$3Ist(b)^-$			−2e			$Co(Ist(b))_3^-$	
−0.39	Co		Mal^{-2}			−2e			$CoMal$	
−0.39	Co		$2Sald^-$			−2e			$Co(Sald)_2$	
−0.39	Co		$3Trop^-$			−2e			$Co(Trop)_3^-$	
−0.39	In		$2Ac^-$			−3e			$In(Ac)_2^+$	
−0.39	In		CNS^-			−3e			$InCNS^{+2}$	

EMF	Elem	Anion	Compound or anion	Cation	H_2O	e =	Cation	Anion	Complex or compound	Compound or element
−0.39	In		$2F^-$			−3e			InF_2^+	
−0.39	In		$3F^-$			−3e			InF_3	
−0.39	Ni		$2AcAc^-$			−2e			$Ni(AcAc)_2$	
−0.39	Ni		$2Af^-$			−2e			$Ni(Af)_2$	
−0.39	Ni		$2Alan^-$			−2e			$Ni(Alan)_2$	
−0.39	Ni		$2Gl^-$			−2e			$Ni(Gl)_2$	
−0.39	Ni		$2Nta^{-3}$			−2e			$Ni(Nta)_2^{-4}$	
−0.39	Ni		OH^-			−2e			$NiOH^+$	
−0.39	Tl		CNS^-			−e			$TlCNS$	
−0.39			$2Coy^-$	UO_2^+		−e			$UO_2(Coy)_2$	
−0.39			$2Hox^-$	UO_2^+		−e			$UO_2(Hox)_2$	
−0.39			$2Mcin^-$	UO_2^+		−e			$UO_2(Mcin)_2$	
−0.39	H		$Neal^-$			−e			$Neal$	
−0.39	H		$4Mxp^-$			−e			$4Mxp$	
−0.39	H		$Ntrm^-$			−e			$Ntrm$	
−0.38	H		HCO_3^-			−e			H_2CO_3	
−0.38	Co		$2Dge^-$			−2e			$Co(Dge)_2$	
−0.38	Co		$Glgl^-$			−2e			$Co(Glgl)^+$	
−0.38	Co		$2Imdp^{-2}$			−2e			$Co(Imdp)_2^{-2}$	
−0.38	Co		$SSald^{-2}$			−2e			$Co(SSald)$	
−0.38	In		$3Ac^-$			−3e			$In(Ac)_3$	
−0.38	In		SO_4^{-2}			−3e			$InSO_4^+$	
−0.38	Ni		$2Dge^-$			−2e			$Ni(Dge)_2$	
−0.38	Ni		$3En$			−2e			$Ni(En)_3^{+2}$	
−0.38	Ni		$Glgl^-$			−2e			$Ni(Glgl)^+$	
−0.38	Ni		$3Pn$			−2e			$Ni(Pn)_3^{+2}$	
−0.38	P				$4H_2O$	−5e	$5H^+$			H_3PO_4
−0.38	Ni		$2Tmen$			−2e			$Ni(Tmen)_2^{+2}$	
−0.38	Pb	$2N_3^-$				−2e				$Pb(N_3)_2$
−0.38	Tl		Cl^-			−e			$TlCl$	
−0.38			$2DmHas^-$	UO_2^+		−e			$UO_2(DmHas)_2$	
−0.38	H		$PhSH^-$			−e			$PhSH$	

EMF	Elem	Anion	Compound or anion	Cation	H_2O	e =	Cation	Anion	Complex or compound	Compound or element
−0.38	H		TMPDA⁻			−e				TMPDA
−0.37	Cd		3CNS⁻			−2e			$Cd(CNS)_3^-$	
−0.37	Co		$2Amac^{-3}$			−2e			$Co(Amac)_2^{-4}$	
−0.37	Co		3En			−2e			$Co(En)_3^{+2}$	
−0.37	Hg		4CN⁻			−2e			$Hg(CN)_4^{-2}$	
−0.37	In		Cl⁻			−3e			$InCl^{+2}$	
−0.37	In		3CNS⁻			−3e			$In(CNS)_3$	
−0.37	Ni		3Ist(b)⁻			−2e			$Ni(Ist(b))_3^-$	
−0.37	Ni		Mal^{-2}			−2e			NiMal	
−0.37	Ni		3Met(b)⁻			−2e			$Ni(Met(b))_3^-$	
−0.37	Ni		2Sald⁻			−2e			$Ni(Sald)_2$	
−0.37	Ni		3Trop⁻			−2e			$Ni(Trop)_3^-$	
−0.37	P				$2H_2O$	−e	H^+			H_3PO_2
−0.37	Pb		2I⁻			−2e			PbI_2	
−0.37	Pb		2Oxd⁻			−2e			$Pb(Oxd)_2$	
−0.37	Pb		2Oxin⁻			−2e			$Pb(Oxin)_2$	
−0.37	Pb		Trop⁻			−2e			$Pb(Trop)^+$	
−0.37	Se		6OH⁻			−4e		SeO_3^{-2}	$3H_2O$	
−0.37				Ti^{+2}		−e	Ti^{+3}			
−0.37	Tl		IO_3^-			−e			$TlIO_3$	
−0.37			2Cin⁻	UO_2^+		−e			$UO_2(Cin)_2$	
−0.36	2Cu		2OH⁻			−2e			Cu_2O	H_2O
−0.36	In		4Ac⁻			−3e			$In(Ac)_4^-$	
−0.36	In		6Ac⁻			−3e			$In(Ac)_6^{-3}$	
−0.36	In		Br⁻			−3e			$InBr^{+2}$	
−0.36	In		2Cl⁻			−3e			$InCl_2^+$	
−0.36	In		3Cl⁻			−3e			$InCl_3$	
−0.36	In		4F⁻			−3e			InF_4^-	
−0.36	Ni		$2Imdp^{-2}$			−2e			$Ni(Imdp)_2^{-2}$	
−0.36	Ni		3Ist(a)⁻			−2e			$Ni(Ist(a))_3^-$	
−0.36	Ni		3Met(a)⁻			−2e			$Ni(Met(a))_3^-$	
−0.36	Ni		$SSald^{-2}$			−2e			$Ni(SSald)$	

EMF	Elem	Anion	Compound or anion	Cation	H_2O	e =	Cation	Anion	Complex or compound	Compound or element
-0.36	Pb		SO_4^{-2}			-2e			$PbSO_4$	
-0.36			3SQ			-e	H^+	$3SQ^-$		
-0.36	H		$4MPy^-$			-e			4MPy	
-0.36	H		$pPhDA^-$			-e			pPhDA	
-0.35			Cd-Hg			-2e	Cd^{+2}			Hg
-0.35	Cd		$6NH_3$			-2e			$Cd(NH_3)_6^{+2}$	
-0.35	Co		$4CNS^-$			-2e			$Co(CNS)_4^{-2}$	
-0.35	Co		$2Glgl^-$			-2e			$Co(Glgl)_2$	
-0.35	Co		$2Ox^{-2}$			-2e			$Co(Ox)_2^{-2}$	
-0.35	Co		SO_4^{-2}			-2e			$CoSO_4$	
-0.35	In		$2Br^-$			-3e			$InBr_2^+$	
-0.35	In		$3Br^-$			-3e			$InBr_3$	
-0.35	In		$2CNS^-$			-3e			$In(CNS)_2^+$	
-0.35	In		I^-			-3e			InI^{+2}	
-0.35	In		$3SO_4^{-2}$			-3e			$In(SO_4)_3^{-3}$	
-0.35	Ni		$2Amac^{-3}$			-2e			$Ni(Amac)_2^{-4}$	
-0.35	Ni		$2Glgl^-$			-2e			$Ni(Glgl)_2$	
-0.35	Ni		Im			-2e			$Ni(Im)^{+2}$	
-0.35	Ni		$P_3O_9^{-3}$			-2e			$NiP_3O_9^-$	
-0.35		SO_4^{-2}	Pb-Hg			-2e			$PbSO_4$	Hg
-0.35	Pb		$2Ist(a)^-$			-2e			$Pb(Ist(a))_2$	
-0.35	Tl		F^-			-e			TlF	
-0.35	Tl		$2P_2O_7^{-4}$			-e			$Tl(P_2O_7)_2^{-7}$	
-0.35	H		$3APy^-$			-e			3APy	
-0.35	H		$NMot^-$			-e			NMot	
-0.35	H		$2MPy^-$			-e			2MPy	
-0.34	Co		NH_3			-2e			$Co(NH_3)^{+2}$	
-0.34	Co		$S_2O_3^{-2}$			-2e			CoS_2O_3	
-0.34	In					-3e	In^{+3}			
-0.34	In		$5Ac^-$			-3e			$In(Ac)_5^{-2}$	
-0.34	In		$2SO_4^{-2}$			-3e			$In(SO_4)_2^-$	
-0.34	Pb		Af^-			-2e			$Pb(Af)^+$	

EMF	Elem	Anion	Compound or anion	Cation	H_2O	e =	Cation	Anion	Complex or compound	Compound or element
-0.34	Pb		$2F^-$			-2e				PbF_2
-0.34	Tl					-e	Tl^+			
-0.34	Tl		OH^-			-e			TlOH (c)	
-0.34			AMSA			-e	H^+	$AMSA^-$		
-0.34	H		Gcd^-			-e			Gcd	
-0.34	H		$3MPy^-$			-e			3MPy	
-0.34	H		$tBPy^-$			-e			tBPy	
-0.34	H		$TAcM^-$			-e			TAcM	
-0.33			$\frac{1}{2}C_2N_2$		H_2O	-e	H^+		HCNO	
-0.33	Co		$2NH_3$			-2e			$Co(NH_3)_2^{+2}$	
-0.33	Co		Pht^{-2}			-2e			Co(Pht)	
-0.33	Ni		$2Im$			-2e			$Ni(Im)_2^{+2}$	
-0.33	Ni		NH_3			-2e			$Ni(NH_3)^{+2}$	
-0.33	Ni		N_2H_4			-2e			$Ni(N_2H_4)^{+2}$	
-0.33	Ni		$2SSald^{-2}$			-2e			$Ni(SSald)_2^{-2}$	
-0.33	Pb		$2Ist(b)^-$			-2e			$Pb(Ist(b))_2$	
-0.33	Pb		$2Met(a)^-$			-2e			$Pb(Met(a))_2$	
-0.33	Pb		$2Met(b)^-$			-2e			$Pb(Met(b))_2$	
-0.33	Pt		H_2S (aq)			-2e	$2H^+$		PtS	
-0.33			$2AcAc^-$	UO_2^+		-e			$UO_2(AcAc)_2$	
-0.33	H		$Acrd^-$			-e			Acrd	
-0.33	H		$4AP^-$			-e			4AP	
-0.33	H		$Bzdl^-$			-e			Bzdl	
-0.33	H		$NiPAl^-$			-e			NiPAl	
-0.32	Ni		$3AcAc^-$			-2e			$Ni(AcAc)_3^-$	
-0.32	Ni		$2NH_3$			-2e			$Ni(NH_3)_2^{+2}$	
-0.32	Ni		$2N_2H_4$			-2e			$Ni(N_2H_4)_2^{+2}$	
-0.32	Ni		SO_4^{-2}			-2e			$NiSO_4$	
-0.32	H		$CNMA^-$			-e			CNMA	
-0.32	H		$DPcA^-$			-e			DPcA	
-0.32	H		$7OHQ$			-e			7OHQ	

EMF	Elem	Anion	Compound or anion	Cation	H_2O	e =	Cation	Anion	Complex or compound	Compound or element
-0.32			Py2C			-e	H^+	Py2C⁻		
-0.31	Ag		$2CN^-$			-e			$Ag(CN)_2^-$	
-0.31	2H		CO_3^{-2}			-2e			H_2CO_3	
-0.31	Co		$3NH_3$			-2e			$Co(NH_3)_3^{+2}$	
-0.31	Ni		$3Im$			-2e			$Ni(Im)_3^{+2}$	
-0.31	Ni		$3N_2H_4$			-2e			$Ni(N_2H_4)_3^{+2}$	
-0.31	Ni		$S_2O_3^{-2}$			-2e			NiS_2O_3	
-0.31	Pb		OH^-			-2e			$PbOH^+$	
-0.31	Pb		$2Trop^-$			-2e			$Pb(Trop)_2$	
-0.31	H		$HRsl^-$			-e			$HRsl$	
-0.31	H		$pExA^-$			-e			$pExA$	
-0.31	H		$5OHQ^-$			-e			$5OHQ$	
-0.31	H		$6OHQ^-$			-e			$6OHQ$	
-0.31	H		$pMxA^-$			-e			$pMxA$	
-0.31	H		Pyr^-			-e			Pyr	
-0.30	Co		$3CNS^-$			-2e			$Co(CNS)_3^-$	
-0.30	Co		$4NH_3$			-2e			$Co(NH_3)_4^{+2}$	
-0.30	Ni		Ac			-2e			$Ni(Ac)^+$	
-0.30	Ni		$4Im$			-2e			$Ni(Im)_4^{+2}$	
-0.30	Ni		$3NH_3$			-2e			$Ni(NH_3)_3^{+2}$	
-0.30	Ni		$4N_2H_4$			-2e			$Ni(N_2H_4)_4^{+2}$	
-0.30	Ni		$5N_2H_4$			-2e			$Ni(N_2H_4)_5^{+2}$	
-0.30	Pb		Cit^{-3}			-2e			$PbCit^-$	
-0.30	Pt		H_2S (g)			-2e	$2H^+$			Pts
-0.30	H		CPC^-			-e			CPC	
-0.30	H		$Nmal^-$			-e			$Nmal$	
-0.30	H		Nea^-			-e			Nea	
-0.30	H		$Oxin^-$			-e			$Oxin$	
-0.30	H		pMa^-			-e			pMa	
-0.30	2H		$18oda^{-2}$			-2e			$18oda$	
-0.30	H		$TMAc^-$			-e			$TMAc$	
-0.29	Co		$5NH_3$			-2e			$Co(NH_3)_5^{+2}$	

EMF	Elem	Anion	Compound or anion	Cation	H_2O	e =	Cation	Anion	Complex or compound	Compound or element
-0.29	Cu		$Data^{-4}$			-2e			$CuData^{-2}$	
-0.29			$2NbO_2$		H_2O	-2e	$2H^+$			Nb_2O_5
-0.29	Ni		$4NH_3$			-2e			$Ni(NH_3)_4^{+2}$	
-0.29	Ni		$6N_2H_4$			-2e			$Ni(N_2H_4)_6^{+2}$	
-0.29	Ni		$2P_2O_7^{-4}$			-2e			$Ni(P_2O_7)_2^{-6}$	
-0.29	Ni		$3Tmen$			-2e			$Ni(Tmen)_3^{+2}$	
-0.29	Pb		$2Af^-$			-2e			$Pb(Af)_2$	
-0.29	Pb		Gl^-			-2e			$Pb(Gl)^+$	
-0.29	Tl		NH_3			-e			$TlNH_3^+$	
-0.29			$2ABA$			-e	H^+	$2ABA^-$		
-0.29			$4ABA$			-e	H^+	$4ABA^-$		
-0.29	H		CHC^-			-e			CHC	
-0.29	H		CPC^-			-e			CPC	
-0.29	H		$DMAc^-$			-e			$DMAc$	
-0.29	H		$DM14NQ^-$			-e			$DM14NQ$	
-0.29	H		$3Mxp^-$			-e			$3Mxp$	
-0.29	H		$NMtAl^-$			-e			$NMtAl$	
-0.29	H		$nOcA^-$			-e			$nOcA$	
-0.29	H		$mPhDA^-$			-e			$mPhDA$	
-0.29	H		Pr^-			-e			Pr	
-0.29			$Py4C$			-e	H^+	$Py4C^-$		
-0.29	H		Qn^-			-e			Qn	
-0.29	H		$nV1A^-$			-e			$nV1A$	
-0.28	Co					-2e	Co^{+2}			
-0.28	Co		CNS^-			-2e			$CoCNS^+$	
-0.28	Co		Nac^-			-2e			$Co(Nac)^+$	
-0.28	Co		Suc^{-2}			-2e			$CoSuc$	
-0.28	Ni		CNS^-			-2e			$NiCNS^+$	
-0.28	Ni		$5Im$			-2e			$Ni(Im)_5^{+2}$	
-0.28	Ni		$2P_4O_{12}^{-4}$			-2e			$Ni(P_4O_{12})_2^{-6}$	
-0.28			$H_3PO_3(aq)$		H_2O	-2e	$2H^+$		$H_3PO_4(aq)$	
-0.28	Pb		$Alan^-$			-2e			$Pb(Alan)^+$	

EMF	Elem	Anion	Compound or anion	Cation	H_2O	e =	Cation	Anion	Complex or compound	Compound or element
-0.28	Pb		$2Br^-$			-2e			$PbBr_2$	
-0.28	Pb		$2S_2O_3^{-2}$			-2e			$Pb(S_2O_3)_2^{-2}$	
-0.28	H		Ac^-			-e			HAc	
-0.28			3ABA			-e	H^+	$3ABA^-$		
-0.28	H		$2AP^-$			-e			2AP	
-0.28	H		$Bzde^-$			-e			Bzde	
-0.28	2H		$14Bda^{-2}$			-2e			14Bda	
-0.28	H		pFA^-			-e			pFA	
-0.28	2H		$pOHBA^{-2}$			-2e			pOHBA	
-0.28	H		mMa^-			-e			mMa	
-0.28	H		$2PeoA^-$			-e			2PeoA	
-0.28			Py3C			-e	H^+	$Py3C^-$		
-0.27	Cu		CNS^-			-e			CuCNS	
-0.27	Cu		2(DNP)			-2e			$Cu(DNP)_2$	
-0.27	Ni		Ac^-			-2e			$Ni(Ac)^+$	
-0.27	Ni		$2Ac^-$			-2e			$Ni(Ac)_2$	
-0.27	Ni		6Im			-2e			$Ni(Im)_6^{+2}$	
-0.27	Ni		$5NH_3$			-2e			$Ni(NH_3)_5^{+2}$	
-0.27	Pb		$2Cl^-$			-2e			$PbCl_2$	
-0.27	H		$ANIL^-$			-e			ANIL	
-0.27	H		$pOHBA^-$			-e			pOHBA	
-0.27	H		$oMxA^-$			-e			oMxA	
-0.27	H		$3PeoA^-$			-e			3PeoA	
-0.27	H		$PSIBP^-$			-e			PSIBP	
-0.27	H		$pPxBA^-$			-e			pPxBA	
-0.27	2H		$mSfP^{-2}$			-2e			mSfP	
-0.26	Co		$2CNS^-$			-2e			$Co(CNS)_2$	
-0.26	Co		$6NH_3$			-2e			$Co(NH_3)_6^{+2}$	
-0.26	Cu		$2OH^-$			-2e			CuO	H_2O
-0.26	Cu		Teta			-2e			$Cu(Teta)^{+2}$	
-0.26	2Hg	$2N_3^-$				-2e				$Hg_2(N_3)_2$
-0.26	Ni		$2CNS^-$			-2e			$Ni(CNS)_2$	

EMF	Elem	Anion	Compound or anion	Cation	H_2O	e =	Cation	Anion	Complex or compound	Compound or element
−0.26	Ni		$3CNS^-$			−2e			$Ni(CNS)_3^-$	
−0.26		$2OH^-$	OH (g)			−e		HO_2^-	H_2O	
−0.26				V^{+2}		−e	V^{+3}			
−0.26	H		$pAxBA^-$			−e				pAxBA
−0.26	H		$Adip^-$			−e				Adip
−0.26	H		$tCnA^-$			−e				tCnA
−0.26	H		$oExA^-$			−e				oExA
−0.26	H		$GltA^-$			−e				GltA
−0.26	H		$2MPa^-$			−e				2MPa
−0.26	H		$pMxBA^-$			−e				pMxBA
−0.26	H		oMa^-			−e				oMa
−0.26	H		$pMBA^-$			−e				pMBA
−0.26	H		$pMTa^-$			−e				pMTa
−0.26	H		$PhAc^-$			−e				PhAc
−0.26	H		$oPhDA^-$			−e				oPhDA
−0.26	2H		$13Pda^{-2}$			−2e				13Pda
−0.26	H		$Pr2C^-$			−e				Pr2C
−0.26	2H		$pSfP^{-2}$			−2e				pSfP
−0.25	Ni					−2e	Ni^{+2}			
−0.25	Ni		Nac^-			−2e			$Ni(Nac)^+$	
−0.25	Ni		$6NH_3$			−2e			$Ni(NH_3)_6^{+2}$	
−0.25		$2OH^-$	OH (aq)			−e		HO_2^-	H_2O	
−0.25	Pb		$2Himda^{-2}$			−2e			$Pb(Himda)_2^{-2}$	
−0.25	Sn		$6F^-$			−4e			SnF_6^{-2}	
−0.25	V				$4H_2O$	−5e	$4H^+$		$V(OH)_4^+$	
−0.25	H		$pAaBA^-$			−e				pAaBA
−0.25	H		$Acrl^-$			−e				Acrl
−0.25	H		$3AP^-$			−e				3AP
−0.25	H		$5Avl^-$			−e				5Avl
−0.25	H		$mBPIP^-$			−e				mBPIP
−0.25	H		$CPICP^-$			−e				CPICP
−0.25	H		$mExA^-$			−e				mExA

EMF	Elem	Anion	Compound or anion	Cation	H_2O	e =	Cation	Anion	Complex or compound	Compound or element
-0.25	H		3OHQ⁻			-e			3OHQ	
-0.25	H		mMxA⁻			-e			mMxA	
-0.25	H		mMBA⁻			-e			mMBA	
-0.25	H		2NptA⁻			-e			2NptA	
-0.25	H		mPhA⁻			-e			mPhA	
-0.25	H		pPhA⁻			-e			pPhA	
-0.25	H		Suc⁻			-e			Suc	
-0.25	H		BA⁻			-e			BA	
-0.24	H		HSe⁻			-e			H_2Se	
-0.24	H		mAaBA⁻			-e			mAaBA	
-0.24	H		mAxBA⁻			-e			mAxBA	
-0.24	H		oAcBA⁻			-e			oAcBA	
-0.24	H		9AQ5DS⁻			-e			9AQ5DS	
-0.24	H		pCa⁻			-e			pCa	
-0.24	H		pCBA⁻			-e			pCBA	
-0.24	H		CNAct⁻			-e			CNAct	
-0.24	H		pFBA⁻			-e			pFBA	
-0.24	H		mOHBA⁻			-e			mOHBA	
-0.24	H		mMxBA⁻			-e			mMxBA	
-0.24	H		oMxBA⁻			-e			oMxBA	
-0.24	H		mMSa⁻			-e			mMSa	
-0.24	H		2NpAm⁻			-e			2NpAm	
-0.24	H		SP3C⁻			-e			SP3C	
-0.23	Cu		2Glu^{-2}			-2e			Cu(Glu)$_2^{-2}$	
-0.23	Pb		2Alan⁻			-2e			Pb(Alan)$_2$	
-0.23	Pb		2Gl⁻			-2e			Pb(Gl)$_2$	
-0.23	Pb		Glgl⁻			-2e			Pb(Glgl)$^+$	
-0.23	3H		PO$_4^{-3}$			-3e			H_3PO_4	
-0.23				$N_2H_5^+$		-4e	5H$^+$			N_2
-0.23	H		mAcBA⁻			-e			mAcBA	
-0.23	H		9AQ26S⁻			-e			9AQ26S	
-0.23	H		9AQ27S⁻			-e			9AQ27S	

EMF	Elem	Anion	Compound or anion	Cation	H_2O	e =	Cation	Anion	Complex or compound	Compound or element
-0.23	H		pBa^-			-e				pBa
-0.23	H		$mBBA^-$			-e				mBBA
-0.23	H		$pBBA^-$			-e				pBBA
-0.23	H		$mCBA^-$			-e				mCBA
-0.23	H		$oCPIP^-$			-e				oCPIP
-0.23	H		$cCnA^-$			-e				cCnA
-0.23	2H		$13D2P^{-2}$			-2e				13D2P
-0.23	H		$26Dma^-$			-e				26Dma
-0.23	H		$mFBA^-$			-e				mFBA
-0.23	H		$F3C^-$			-e				F3C
-0.23	H		$OHAc^-$			-e				OHAc
-0.23	H		$mIBA^-$			-e				mIBA
-0.23	H		$pIBA^-$			-e				pIBA
-0.23	H		$oMBA^-$			-e				oMBA
-0.23	H		$1NpAm^-$			-e				1NpAm
-0.23	H		$mPxBA^-$			-e				mPxBA
-0.22	Cu		NH_3			-2e			$Cu(NH_3)^{+2}$	
-0.22	Cu		$Edta^{-4}$			-2e			$CuEdta^{-2}$	
-0.22	Cu	$2OH^-$				-2e				$Cu(OH)_2$
-0.22	Cu		Tate			-2e			$Cu(Tate)^{+2}$	
-0.22	Pb		Dyp			-2e			$Pb(Dyp)^{+2}$	
-0.22			$S_2O_6^{-2}$		$2H_2O$	-2e	$4H^+$	$2SO_4^{-2}$		
-0.22			F^-	UO_2^+		-e			UO_2F^+	
-0.22	H		$pAcBA^-$			-e				pAcBA
-0.22			A3SA			-e	H^+		$A3SA^-$	
-0.22	H		$BdGr^-$			-e				BdGr
-0.22	H		$CbAc^-$			-e				CbAc
-0.22	H		$CNAd^-$			-e				CNAd
-0.22	H		$26CPIP^-$			-e				26CPIP
-0.22	H		$COOH^-$			-e				HCOOH
-0.22	H		pIa^-			-e				pIa
-0.22	H		$mMxCa^-$			-e				mMxCa

EMF	Elem	Anion	Compound or anion	Cation	H_2O	e =	Cation	Anion	Complex or compound	Compound or element
-0.22	H		MSAc⁻			-e				MSAc
-0.22	H		1NptA⁻			-e				1NptA
-0.22	H		oPha⁻			-e				oPha
-0.22	H		2tBa⁻			-e				2tBa
-0.22	H		TGa⁻			-e				TGa
-0.21	2H		HPO_4^{-2}			-2e				H_3PO_4
-0.21	Pb		2Glgl⁻			-2e				$Pb(Glgl)_2$
-0.21	2H		SO_3^{-2}			-2e				H_2SO_3
-0.21	H		oAaBA⁻			-e				oAaBA
-0.21	H		oAxBA⁻			-e				oAxBA
-0.21	H		AcAcA⁻			-e				AcAcA
-0.21	H		3APA⁻			-e				3APA
-0.21	H		9AQDS⁻			-e				9AQDS
-0.21	H		mBa⁻			-e				mBa
-0.21	H		mCNBA⁻			-e				mCNBA
-0.21	H		pCNBA⁻			-e				pCNBA
-0.21	2H		$12Eda^{-2}$			-2e				12Eda
-0.21	H		mFa⁻			-e				mFa
-0.21	H		mIa⁻			-e				mIa
-0.21	H		MxAc⁻			-e				MxAc
-0.21	H		mNOBA⁻			-e				mNOBA
-0.21	H		oPxBA⁻			-e				oPxBA
-0.21	2H		$PhPhA^{-2}$			-2e				PhPhA
-0.21	H		iPHA⁻			-e				iPhA
-0.21	H		Phz⁻			-e				Phz
-0.21	H		mSmBA⁻			-e				mSmBA
-0.21	H		pSmBA⁻			-e				pSmBA
-0.21	H		TPhA⁻			-e				TPhA
-0.21	H		SP2C⁻			-e				SP2C
-0.21	H		mTFMa⁻			-e				mTFMa
-0.20			HCOOH(aq)			-2e	$2H^+$			$CO_2(g)$
-0.20	Mo					-3e	Mo^{+3}			

EMF	Elem	Anion	Compound or anion	Cation	H_2O	e =	Cation	Anion	Complex or compound	Compound or element
-0.20	H		NO_2^-			-e				HNO_2
-0.20	H		$9AQ1S^-$			-e				$9AQ1S$
-0.20	H		mCa^-			-e				mCa
-0.20	H		$ExCAc^-$			-e				$ExCAc$
-0.20	H		$pNOBA^-$			-e				$pNOBA$
-0.20	H		$oPhBA^-$			-e				$oPhBA$
-0.19	2H		CrO_4^{-2}			-2e				H_2CrO_4
-0.19	Cu		I^-			-e				CuI
-0.19	H		F^-			-e				HF
-0.19	2In				$3H_2O$	-6e	$6H^+$			In_2O_3
-0.19	Pb		Ac^-			-2e				$PbAc^+$
-0.19	H		AcH_2^-			-e				AcH_2
-0.19	H		$3AcPy^-$			-e				$3AcPy$
-0.19	H		$9AQ2S^-$			-e				$9AQ2S$
-0.19	H		Cit^-			-e				Cit
-0.19	2H		$12CHDA^{-2}$			-2e				$12CHDA$
-0.19	H		oFa^-			-e				oFa
-0.19	H		$oFBA^-$			-e				$oFBA$
-0.19	H		$F2C^-$			-e				$F2C$
-0.19	H		IAc^-			-e				IAc
-0.19	H		$2MxPy^-$			-e				$2MxPy$
-0.19	2H		$MePhs^{-2}$			-2e				$MePhs$
-0.18	H		F^-			-e				HF
-0.18	H		HTe^-			-e				H_2Te
-0.18	Pb		Cl^-			-2e				$PbCl^+$
-0.18	2H		HPO_3^{-2}			-2e				H_3PO_3
-0.18	H		$3ABA^-$			-e				$3ABA$
-0.18			$A4SA$			-e	H^+			$A4SA^-$
-0.18	H		$DCPIC^-$			-e				$DCPIC$
-0.18	H		$FumA^-$			-e				$FumA$
-0.18	2H		$MlcA^{-2}$			-2e				$MlcA$

EMF	Elem	Anion	Compound or anion	Cation	H_2O	e =	Cation	Anion	Complex or compound	Compound or element
-0.18	H		Sal^-			-e			Sal	
-0.18	H		$Tart^-$			-e			Tart	
-0.17	Cu		Hed^{-3}			-2e			$Cu(Hed)^-$	
-0.17	Fe		$Edta^{-4}$			-3e			$FeEdta^-$	
-0.17	In				$3H_2O$	-3e	$3H^+$			$In(OH)_3$
-0.17	Pb		$3S_2O_3^{-2}$			-2e			$Pb(S_2O_3)_3^{-4}$	
-0.17	Sn		Cl^-			-2e			$SnCl^+$	
-0.17	H		$BrAc^-$			-e			BrAc	
-0.17	H		$oBBA^-$			-e			oBBA	
-0.17	H		$ClAc^-$			-e			ClAc	
-0.17	H		$oCBA^-$			-e			oCBA	
-0.17	H		$3CPy^-$			-e			3CPy	
-0.17	H		$oIBA^-$			-e			oIBA	
-0.17	H		Mal^-			-e			Mal	
-0.17	2H		Mal^{-2}			-2e			Mal	
-0.17	H		$oPha^-$			-2e			oPha	
-0.17	2H		$Pipz^{-2}$			-2e			Pipz	
-0.17	2H		Suc^{-2}			-2e			Suc	
-0.16	Pb		Br^-			-2e			$PbBr^+$	
-0.16	Cu		2(13DAP)			-2e			$Cu(13DAP)_2^{+2}$	
-0.16	Sn		Br^-			-2e			$SnBr^+$	
-0.16	Sn		$2Cl^-$			-2e			$SnCl_2$	
-0.16	2H		$Adip^{-2}$			-2e			Adip	
-0.16	H		oCa^-			-e			oCa	
-0.16	H		$mCNa^-$			-e			mCNa	
-0.16	2H		$Glta^{-2}$			-2e			Glta	
-0.16	H		$mMSfa^-$			-e			mMSfa	
-0.16	2H		$oPhA^{-2}$			-2e			oPhA	
-0.15	Ag		I^-			-e			AgI	
-0.15	Ge				$2H_2O$	-4e	$4H^+$			GeO_2
-0.15	Ru	$4OH^-$				-4e				$Ru(OH)_4$
-0.15	Sn		$2Br^-$			-2e			$SnBr_2$	

EMF	Elem	Anion	Compound or anion	Cation	H_2O	e =	Cation	Anion	Complex or compound	Compound or element
-0.15	Sn		$3Br^-$			-2e			$SnBr_3^-$	
-0.15			$2F^-$	UO_2^+		-e			UO_2F_2	
-0.15	H		oBa^-			-e			oBa	
-0.15	H		$ByoA^-$			-e			ByoA	
-0.15	H		$CNAc^-$			-e			CNAc	
-0.15	H		FAc^-			-e			FAc	
-0.15	H		oIa^-			-e			oIa	
-0.15	H		$mNOa^-$			-e			mNOa	
-0.15	H		$OxAc^-$			-e			OxAc	
-0.15	H		$Pyzl^-$			-e			Pyzl	
-0.15	H		$Thzl^-$			-e			Thzl	
-0.15	H		$SCNAc^-$			-e			SCNAc	
-0.15	H		$pTFMa^-$			-e			pTFMa	
-0.14	H		AsO_3^-			-e			$HAsO_3$	
-0.14			B_2H_6			-6e	$6H^+$			2B
-0.14	Pb		$2Ac^-$			-2e			$Pb(Ac)_2$	
-0.14	Pb		NO_3^-			-2e			$PbNO_3^+$	
-0.14	3H		$HP_2O_7^{-3}$			-3e			$H_4P_2O_7$	
-0.14	In					-e	In^+			
-0.14	Sn					-2e	Sn^{+2}			
-0.14	Sn		$3Cl^-$			-2e			$SnCl_3^-$	
-0.14	H		$4ABA^-$			-e			4ABA	
-0.14	2H		Cit^{-2}			-2e			Cit	
-0.14	H		$3SQ^-$			-e			3SQ	
-0.14	H		$pMxCa^-$			-e			pMxCa	
-0.14	H		$MSOAc^-$			-e			MSOAc	
-0.14	2H		$iPhA^{-2}$			-2e			iPhA	
-0.14	H		$Pydz^-$			-e			Pydz	
-0.13		$5OH^-$	$Cr(OH)_3$			-3e		CrO_4^{-2}	$4H_2O$	
-0.13	Cu		Deta			-2e			$Cu(Deta)^{+2}$	
-0.13	2Cu	$(FeCy)^{-4}$				-4e			$Cu_2(FeCy)$	
-0.13			HO_2			-e	H^+			O_2

EMF	Elem	Anion	Compound or anion	Cation	H_2O	e =	Cation	Anion	Complex or compound	Compound or element
-0.13	4H		$P_2O_7^{-4}$			-4e			$H_4P_2O_7$	
-0.13	Pb					-2e	Pb^{+2}			
-0.13	Pb		$3Ac^-$			-2e			$Pb(Ac)_3^-$	
-0.13	Pb		Nac^-			-2e			$Pb(Nac)^+$	
-0.13	H		$2AAc^-$			-e			$2AAc$	
-0.13	H		$2Apa^-$			-e			$2Apa$	
-0.13	3H		Cit^{-3}			-3e			Cit	
-0.13	2H		$FumA^{-2}$			-2e			$FumA$	
-0.13	H		$4OHQ^-$			-e			$4OHQ$	
-0.13	H		$oMxCa^-$			-e			$oMxCa$	
-0.13	H		$oNOBA^-$			-e			$oNOBA$	
-0.13	2H		Ox^{-2}			-2e			Ox	
-0.13	2H		$OxAc^{-2}$			-2e			$OxAc$	
-0.13	2H		$Tart^{-2}$			-2e			$Tart$	
-0.13	2H		$TPhA^{-2}$			-2e			$TPhA$	
-0.12	H		ClO_2^-			-e			$HClO_2$	
-0.12	Cu		$2En$			-e			$Cu(En)_2^+$	
-0.12	Cu		$2NH_3$			e			$Cu(NH_3)_2^+$	
-0.12	H		$H_2PO_4^-$			-e			H_3PO_4	
-0.12	Os	$4OH^-$				-4e				$Os(OH)_4$
-0.12	Pb		$4Ac^-$			-2e			$Pb(Ac)_4^{-2}$	
-0.12	W				$2H_2O$	-4e	$4H^+$			WO_2
-0.12	H		$2ABA^-$			-e			$2ABA$	
-0.12	H		$NSIP^-$			-e			$NSIP$	
-0.12	H		$NSICP^-$			-e			$NSICP$	
-0.12	H		$Py3C^-$			-e			$Py3C$	
-0.12	H		$T1B1^-$			-e			$T1B1$	
-0.11	H		$H_2PO_3^-$			-e			H_3PO_3	
-0.11	H		HSO_3^-			-e			H_2SO_3	
-0.11	Sn				$2H_2O$	-4e	$4H^+$			SnO_2
-0.11	2H		Bzd^{-2}			-2e			Bzd	
-0.11	H		$2IPy^-$			-e			$2IPy$	

EMF	Elem	Anion	Compound or anion	Cation	H_2O	e =	Cation	Anion	Complex or compound	Compound or element
−0.11	H		$MlcA^-$			−e			$MlcA$	
−0.11	H		$PhPhA^-$			−e			$PhPhA$	
−0.11	H		$PplA^-$			−e			$PplA$	
−0.11	2H		$mSfBA^{-2}$			−2e			$mSfBA$	
−0.11	2H		$pSfBA^{-2}$			−2e			$pSfBA$	
−0.10			$3F^-$	UO_2^+		−e			$UO_2F_3^-$	
−0.10	Sn				H_2O	−2e	$2H^+$			SnO
−0.10	H		$pCNa^-$			−e			$pCNa$	
−0.10	H		$NOAc^-$			−e			$NOAc$	
−0.10	2H		$pPhDA^{-2}$			−2e			$pPhDA$	
−0.10	H		$Py4C^-$			−e			$Py4C$	
−0.10	H		$pTSiA^-$			−e			$pTSiA$	
−0.09	Cu		$S_2O_3^{-2}$			−e			$CuS_2O_3^-$	
−0.09	H		IO_4^-			−e			HIO_4	
−0.09			Gly^-	UO_2^+		−e			$UO_2(Gly)^+$	
−0.09	W				$3H_2O$	−6e	$6H^+$		WO_3	
−0.09	H		$AESA^-$			−e			$AESA$	
−0.09	H		$BSiA^-$			−e			$BSiA$	
−0.09	H		$Bztrz^-$			−e			$Bztrz$	
−0.09	H		$3CNPy^-$			−e			$3CNPy$	
−0.09	H		$MePhs^-$			−e			$MePhs$	
−0.09	H		$pMSfa^-$			−e			$pMSfa$	
−0.08		$2OH^-$	Cu_2O		H_2O	−2e			$2Cu(OH)_2$	
−0.08	Cu		Cit^{-3}			−2e			$Cu(Cit)^-$	
−0.08	Cu	$2IO_3^-$				−2e				$Cu(IO_3)_2$
−0.08		HO_2^-	OH^-			−2e			H_2O	O_2
−0.08			$H_2SO_4^-$		$2H_2O$	−2e	H^+		$2H_2SO_3$	
−0.08	H		$MelA^-$			−e			$MelA$	
−0.08	H		Ox^-			−e			Ox	
−0.08	2H		$mPhDA^{-2}$			−2e			$mPhDA$	
−0.08	H		$Pymd^-$			−e			$Pymd$	
−0.07	H		$H_2PO_2^-$			−e			H_3PO_2	

EMF	Elem	Anion	Compound or anion	Cation	H_2O	e =	Cation	Anion	Complex or compound	Compound or element
-0.07	Mo				$2H_2O$	-4e	$4H^+$			MoO_2
-0.07	H		SO_4^{-2}			-e		HSO_4^-		
-0.07	H		Cna^-			-e			Cna	
-0.07	H		$DClAc^-$			-e			DClAc	
-0.07	3H		$MelA^{-3}$			-3e			MelA	
-0.07	4H		$MelA^{-4}$			-4e			MelA	
-0.07	5H		$MelA^{-5}$			-5e			MelA	
-0.07	6H		$MelA^{-6}$			-6e			MelA	
-0.07	H		$Phnz^-$			-e			Phnz	
-0.07	2H		$TMPhDA^{-2}$			-2e			TMPhDA	
-0.06	Cu		$mOxin^-$			-2e			$Cu(mOxin)^+$	
-0.06	Cu		$Oxin^-$			-2e			$Cu(Oxin)^+$	
-0.06			PH_3 (g)			-3e	$3H^+$			P(white)
-0.06	H		$H_3P_2O_7^-$			-e			$H_4P_2O_7$	
-0.06	2H		$H_2P_2O_7^{-2}$			-2e			$H_4P_2O_7$	
-0.06	2H		SeO_4^{-2}			-2e			H_2SeO_4	
-0.06	2H		SO_4^{-2}			-2e			H_2SO_4	
-0.06	H		$DCna^-$			-e			DCna	
-0.06	2H		$MelA^{-2}$			-2e			MelA	
-0.06	H		$pNOa^-$			-e			pNOa	
-0.06	2H		$oPhDA^{-2}$			-2e			oPhDA	
-0.06	H		$Py2C^-$			-e			Py2C	
-0.05	H		IO_3^-			-e			HIO_3	
-0.05		$2OH^-$	$Mn(OH)_2$			-2e			MnO_2	$2H_2O$
-0.05		$2OH^-$	TlOH			-2e			$Tl(OH)_3$	
-0.05			SO_4^{-2}	UO_2^+		-e			UO_2SO_4	
-0.05	H		$2BrPy^-$			-e			2BrPy	
-0.05	H		$CsBl^-$			-e			CsBl	
-0.05	H		$DPhAm^-$			-e			DPhAm	
-0.05	H		$4SQ^-$			-e			4SQ	
-0.05	H		$PyNO^-$			-e			PyNO	
-0.04	Cu		$2Oxin^-$			-2e			$Cu(Oxin)_2$	

EMF	Elem	Anion	Compound or anion	Cation	H_2O	e =	Cation	Anion	Complex or compound	Compound or element
-0.04	Fe					-3e	Fe^{+3}			
-0.04	2Hg		$2I^-$			-2e			Hg_2I_2	
-0.04	Hg		$4I^-$			-2e			HgI_4^{-2}	
-0.04			$2Gly^-$	UO_2^+		-e			$UO_2(Gly)_2$	
-0.04	H		$2CPy^-$			-e			2CPy	
-0.04	H		$Pyzn^-$			-e			Pyzn	
-0.04	H		$Qnxl^-$			-e			Qnxl	
-0.04	H		$TClAc^-$			-e			TClAc	
-0.04	H		$TNPh^-$			-e			TNPh	
-0.04	H		$TNOBA^-$			-e			TNOBA	
-0.03	Cu		$2mOxin^-$			-e			$Cu(mOxin)_2$	
-0.03			$4F^-$	UO_2^+		-e			$UO_2F_4^{-2}$	
-0.03	H		$A4SA^-$			-e			A4SA	
-0.02	Ag		CN^-			-e			AgCN	
-0.02	Cu		Glu^{-2}			-2e			Cu(Glu)	
-0.02			$3Gly^-$	UO_2^+		-e			$UO_2(Gly)_3^-$	
-0.02	Cu	$2S_2O_3^{-2}$				-2e			$Cu(S_2O_3)_2^{-2}$	
-0.02	H		$Actld^-$			-e			Actld	
-0.02	Cu	$4SCN^-$				-e			$Cu(SCN)_4^{-3}$	
-0.02	H		$A3SA^-$			-e			A3SA	
-0.02	H		$26M4P^-$			-e			26M4P	
-0.02	H		$Glcn^-$			-e			Glcn	
-0.01	Cu		$Ndap^{-3}$			-2e			$Cu(Ndap)^-$	
-0.01	H		$MeBl^-$			-e			MeBl	
-0.01	H		$TFAc^-$			-e			TFAc	
-0.01	H		Ur^-			-e			Ur	
-0.01	2H		Ur^{-2}			-2e			Ur	
0.00	At_2		$4OH^-$			-2e		$2AtO^-$		$2H_2O$
0.00				Ge^{+2}		-2e	Ge^{+4}			
0.00	D_2					-2e	$2D^+$			
0.00	Pd		$4NH_3$			-2e			$Pd(NH_3)_4^{+2}$	
0.00	H_2					-2e	$2H^+$			

EMF	Elem	Anion	Compound or anion	Cation	H_2O	e =	Cation	Anion	Complex or compound	Compound or element
0.00	Rh	$3OH^-$				-3e				$Rh(OH)_3$
0.00			$3SO_4^{-2}$	UO_2^+		-e			$UO_2(SO_4)_3^{-4}$	
0.00				Ti^{+3}		-e	Ti^{+4}			
0.00	3H		Ur^{-3}			-3e				Ur
+0.01	Cu		Ptn			-2e			$Cu(Ptn)^{+2}$	
+0.01	Fe		Sal^{-2}			-3e			$Fe(Sal)^+$	
+0.01		$2OH^-$	NO_2^-			-2e	NO_3^-			H_2O
+0.01			CNS^-	UO_2^+		-e			UO_2CNS^+	
+0.01			$2SO_4^{-2}$	UO_2^+		-e			$UO_2(SO_4)_2^{-2}$	
+0.01	2H		$Actd^{-2}$			-2e			Actd	
+0.02	Ag		$2S_2O_3^{-2}$			-e			$Ag(S_2O_3)_2^{-3}$	
+0.02	Cu		DNP^-			-2e			$CuDNP^+$	
+0.02	Os		$9OH^-$			-8e			$HOsO_5$	$4H_2O$
+0.02			$3CNS^-$	UO_2^+		-e			$UO_2(CNS)_3^-$	
+0.02	H		$2OHQ^-$			-e			2OHQ	
+0.02	H		$oNOa^-$			-e			oNOa	
+0.02	H		$Pyrr^-$			-e			Pyrr	
+0.03	Cu		Br^-			+e			CuBr	
+0.03	Cu	N_3^-				-e				CuN_3
+0.03	Cu		En			-2e			$Cu(En)^{+2}$	
+0.03	Cu		Has^-			-2e			$Cu(Has)^+$	
+0.03	Cu		$3NH_3$			-2e			$Cu(NH_3)_3^{+2}$	
+0.03	Cu		$Imda^{-2}$			-2e			Cu(Imda)	
+0.03	Cu		$Impa^{-2}$			-2e			Cu(Impa)	
+0.03	Cu		Pn			-2e			$Cu(Pn)^{+2}$	
+0.03	Cu		Sal^{-2}			-2e			Cu(Sal)	
+0.03	Hg	$2S_2O_3^{-2}$				-2e			$Hg(S_2O_3)_2^{-2}$	
+0.03	H		$Actd^-$			-e				Actd
+0.03	Hg		$4SeCN^-$			-2e			$Hg(SeCN)_4^{-2}$	
+0.03	H		$2FPy^-$			-e			2FPy	
+0.04	Cu		$DmHas^-$			-2e			$Cu(DmHas)^+$	
+0.04	Cu		Nta^{-3}			-2e			$Cu(Nta)^-$	

EMF	Elem	Anion	Compound or anion	Cation	H_2O	e =	Cation	Anion	Complex or compound	Compound or element
+0.04	Cu		Oxd^-			-2e			$Cu(Oxd)^+$	
+0.04	Cu		Tmen			-2e			$Cu(Tmen)^{+2}$	
+0.04	2Rh		$6OH^-$			-6e			Rh_2O_5	$3H_2O$
+0.04	H		CSS^-			-e			CSS	
+0.05	Cu		$2DmHas^-$			-2e			$Cu(DmHas)_2$	
+0.05	Cu		13DAP			-2e			$Cu(13DAP)^{+2}$	
+0.05	Cu		Hox			-2e			$Cu(Hox)^{+2}$	
+0.05		$2OH^-$	SeO_3^{-2}			-2e			SeO_4^{-2}	H_2O
+0.05			Pd_2H			-e	H^+			2Pd
+0.05				UO_2^+		-e	UO_2^{+2}			
+0.05			$2CNS^-$	UO_2^+		-e			$UO_2(CNS)_2$	
+0.05	Zn				$2H_2O$	-2e	$3H^+$	$HZnO_2^-$		
+0.06	Cd				H_2O	-2e	$2H^+$			CdO
+0.06	Cu		Cin^-			-2e			$Cu(Cin)^+$	
+0.06	Cu		Coy^-			-2e			$Cu(Coy)^+$	
+0.06	Cu		$2Has^-$			-2e			$Cu(Has)_2$	
+0.06	Cu		$Imdp^{-2}$			-2e			$Cu(Imdp)$	
+0.06	Cu		$2Oxd^-$			-2e			$Cu(Oxd)_2$	
+0.06	H		ClO_3^-			-e			$HClO_3$	
+0.06	H		$HCrO_4^-$			-e			H_2CrO_4	
+0.06			HCHO (aq)		H_2O	-2e	$2H^+$		HCOOH (aq)	
+0.06			$2NH_3$			-6e	$6H^+$			N_2
+0.06	H		$MCB1^-$			-e			MCB1	
+0.06	H		TUr^-			-e			TUr	
+0.07	Ag		Br^-			-e			AgBr	
+0.07	Cu		Af^-			-2e			$Cu(Af)^+$	
+0.07	Cu		2En			-2e			$Cu(En)_2^{+2}$	
+0.07	Cu		Ntp^{-3}			-2e			$Cu(Ntp)^-$	
+0.07	Cu		2Pn			-2e			$Cu(Pn)_2^{+2}$	
+0.07	Cu		2Ptn			-2e			$Cu(Ptn)_2^{+2}$	
+0.07	Pd		$2OH^-$			-2e			$Pd(OH)_2$	

EMF	Elem	Anion	Compound or anion	Cation	H_2O	e =	Cation	Anion	Complex or compound	Compound or element
+0.07			Br^-	UO_2^+		-e			UO_2Br^+	
+0.07			Cl^-	UO_2^+		-e			UO_2Cl^+	
+0.08	Cu		$2Hox^-$			-2e			$Cu(Hox)_2$	
+0.08	Cu		$MpHas^-$			-2e			$Cu(MpHas)^+$	
+0.08	Cu		SO_3^{-2}			-e			$CuSO_3^-$	
+0.08			$2S_2O_3^{-2}$			-2e		$S_4O_6^{-2}$		
+0.08	H		ITS^-			-e			ITS	
+0.09	Cu		$AcAc^-$			-2e			$Cu(AcAc)^+$	
+0.09	Cu		$2Af^-$			-2e			$Cu(Af)_2$	
+0.09	Cu		$Alan^-$			-2e			$Cu(Alan)^+$	
+0.09	Cu		$Aspa^{-2}$			-2e			$Cu(Aspa)$	
+0.09	Cu		$2Cin^-$			-2e			$Cu(Cin)_2$	
+0.09	Cu		Gl^-			-2e			$Cu(Gl)^+$	
+0.09	Cu		$2MpHas^-$			-2e			$Cu(MpHas)_2$	
+0.09			NO_3^-	UO_2^+		-e			$UO_2NO_3^+$	
+0.09	Pt	$4CN^-$				-2e			$Pt(CN)_4^{-2}$	
+0.09	H		$2SQ^-$			-e			$2SQ$	
+0.10	Cu		Dge			-2e			$Cu(Dge)^+$	
+0.10	Fe		OH^-			-3e			$FeOH^{+2}$	
+0.10	Fe		$2Sal^{-2}$			-3e			$Fe(Sal)_2^-$	
+0.10	Hg		$2OH^-$			-2e			HgO (r)	H_2O
+0.10	Ir		$6OH^-$			-6e			Ir_2O_3	$3H_2O$
+0.10	H		NO_3^-			-e			HNO_3	
+0.10		$4OH^-$	$Pt(OH)_2$			-2e			$Pt(OH)_6^{-2}$	
+0.10			SiH_4 (g)			-4e	$4H^+$			Si
+0.10				Ti^{+3}	H_2O	-e	$2H^+$		TiO^{+2}	
+0.11			$Co(NH_3)_6^{+2}$			-e			$Co(NH_3)_6^{+3}$	
+0.11	Cu		$2Trop^-$			-2e			$Cu(Trop)_2$	
+0.11		$2OH^-$	$2NH_4OH$			-2e			N_2H_4	$4H_2O$
+0.12	Bi		Br^-			-3e			$BiBr^{+2}$	
+0.12	Cu		$Sald^-$			-2e			$Cu(Sald)^+$	
+0.12	Fe		$2OH^-$			-3e			$Fe(OH)_2^+$	

EMF	Elem	Anion	Compound or anion	Cation	H_2O	e =	Cation	Anion	Complex or compound	Compound or element
+0.12	Ge					-4e	Ge^{+4}			
+0.13			CH_4 (g)			-4e	$4H^+$			C
+0.13	Cu		$2Coy^-$			-2e			$Cu(Coy)_2$	
+0.13	Cu		2Tmen			-2e			$Cu(Tmen)_2^{+2}$	
+0.13	Mn		Ox^{-2}			-3e			$Mn(Ox)^+$	
+0.13		Re^-				-4e	Re^{+3}			
+0.13	H		IDS^-			-e			IDS	
+0.14	Cu		$2AcAc^-$			-2e			$Cu(AcAc)_2$	
+0.14	Cu		$2Alan^-$			-2e			$Cu(Alan)_2$	
+0.14	Cu		$2Aspa^{-2}$			-2e			$Cu(Aspa)_2^{-2}$	
+0.14	Cu		Cl^-			-e			CuCl	
+0.14	Cu		$2Gl^-$			-2e			$Cu(Gl)_2$	
+0.14	Cu		$P_2O_7^{-4}$			-2e			$CuP_2O_7^{-2}$	
+0.14	Fe		$AcAc^-$			-3e			$Fe(AcAc)^{+2}$	
+0.14	Fe		Ox^{-2}			-3e			$Fe(Ox)^+$	
+0.14			H_2S (aq)			-2e	$2H^+$			S
+0.14	2Hg		$2Br^-$			-2e			Hg_2Br_2	
+0.14	Pd	$4SCN^-$				-2e			$Pd(SCN)_4^{-2}$	
+0.14	H		MnO_4^-			-e			$HMnO_4$	
+0.14	H		$Glpn^-$			-e			Glpn	
+0.15	4Ag	$(FeCy)^{-4}$				-4e			$Ag_4(FeCy)$	
+0.15	Bi		Cl^-			-3e			$BiCl^{+2}$	
+0.15				Cu^+		-e	Cu^{+2}			
+0.15	Cu		NH_3			-e			$CuNH_3^+$	
+0.15	Cu		OH^-			-2e			$CuOH^+$	
+0.15	Cu		$2DMMal^{-2}$			-2e			$Cu(DMMal)_2^{-2}$	
+0.15	Cu		$2Sal^{-2}$			-2e			$Cu(Sal)_2^{-2}$	
+0.15	Cu		Tf^-			-2e			$Cu(Tf)^+$	
+0.15	Fe		$2AcAc^-$			-3e			$Fe(AcAc)_2^+$	
+0.15		OH^-	$Mn(OH)_2$			-e			$Mn(OH)_3$	
+0.15	Mo				$4H_2O$	-6e	$8H^+$	MoO_4^{-2}		
+0.15				Np^{+3}		-e	Np^{+4}			

EMF	Elem	Anion	Compound or anion	Cation	H_2O	e =	Cation	Anion	Complex or compound	Compound or element
+0.15	Pt		$2OH^-$			-2e			$Pt(OH)_2$	
+0.15	2Sb				$3H_2O$	-6e	$6H^+$		Sb_2O_3	
+0.15				Sn^{+2}		-2e	Sn^{+4}			
+0.16	Bi		Cl^-		H_2O	-3e	$2H^+$		$BiOCl$	
+0.16	Bi		NO_3^-			-3e			$BiNO_3^{+2}$	
+0.16	Cu		$Glgl^-$			-2e			$Cu(Glgl)^+$	
+0.16	Cu		Ox^{-2}			-2e			$Cu(Ox)$	
+0.17		OH^-	$Co(OH)_2$			-e			$Cu(OH)_3$	
+0.17	Cu		$2Br^-$			-e			$CuBr_2^-$	
+0.17	Cu		$2Glgl^-$			-2e			$Cu(Glgl)_2$	
+0.17	Cu	HPO_4^-				-2e				$CuHPO_4$
+0.17	Cu		$2Imda^{-2}$			-2e			$Cu(Imda)_2^{-2}$	
+0.17	Cu		Mal^{-2}			-2e			$Cu(Mal)$	
+0.17	Cu		$2Sald^-$			-2e			$Cu(Sald)_2$	
+0.17	Fe		$2Nta^{-3}$			-3e			$Cu(Nta)_2^{-3}$	
+0.17			H_2SO_3		H_2O	-2e	$4H^+$	SO_4^{-2}		
+0.17	Sb	$4Cl^-$				-3e			$SbCl_4^-$	
+0.17	H		$BAzB^-$			-e			$BAzB$	
+0.17	H		Dox^-			-e			Dox	
+0.18	Bi		$2Br^-$			-3e			$BiBr_2^+$	
+0.18	Bi		CNS^-			-3e			$Bi(CNS)^{+2}$	
+0.18	Bi		$2CNS^-$			-3e			$Bi(CNS)_2^+$	
+0.18	Cu		$2Deta$			-2e			$Cu(Deta)_2^{+2}$	
+0.18	Cu		$SSald^{-2}$			-2e			$Cu(SSald)$	
+0.18	Fe		$3AcAc^-$			-3e			$Fe(AcAc)_3$	
+0.18	Pd	$4I^-$				-2e		PdI_4^{-2}		
+0.18	H		$HSeO_4^-$			-e			H_2SeO_4	
+0.18	H		HSO_4^-			-e			H_2SO_4	
+0.19	Bi		$3Br^-$			-3e			$BiBr_3$	
+0.19	Bi		$3Cl^-$			-3e			$BiCl_3$	
+0.19	Bi		$2Cl^-$			-3e			$BiCl_2^+$	
+0.19	Cu		$2Dge^-$			-2e			$Cu(Dge)_2$	

EMF	Elem	Anion	Compound or anion	Cation	H_2O	e =	Cation	Anion	Complex or compound	Compound or element
+0.19	Cu		$EMal^{-2}$			-2e				$Cu(EMal)$
+0.19	Fe		Tf^-			-3e			$Fe(Tf)^{+2}$	
+0.20			$2At^-$			-2e				At_2
+0.20	Bi		$4Cl^-$			-3e			$BiCl_4^-$	
+0.20	Bi					-3e	Bi^{+3}			
+0.20	Cu	$2H_2PO_4^-$				-2e			$Cu(H_2PO_4)_2$	
+0.20	Cu		$DMMal^{-2}$			-2e			$Cu(DMMal)$	
+0.20	Fe		$2Ox^{-2}$			-3e			$Fe(Ox)_2^-$	
+0.20	Mn		$2Ox^{-2}$			-3e			$Mn(Ox)_2^-$	
+0.20			Cu_2O	$2H^+$		-2e	$2Cu^{+2}$			H_2O
+0.21	C				$2H_2O$	-4e	$4H^+$			CO_2
+0.21	Cu		$2Himda^{-2}$			-2e			$Cu(Himda)_2^{-2}$	
+0.21	Cu		Im			-2e			$Cu(Im)^{+2}$	
+0.21	Cu		$2Impa^{-2}$			-2e			$Cu(Impa)_2^{-2}$	
+0.21			$(RuNh)^{+2}$			-e			$(RuNh)^{+3}$	
+0.21	H		Dee^-			-e			Dee	
+0.22	Cu		Bu^-			-2e			$Cu(Bu)^+$	
+0.22	Cu		$2SSald^{-2}$			-2e			$Cu(SSald)_2^{-2}$	
+0.22	Fe		$3Sal^{-2}$			-3e			$Fe(Sal)_3^{-3}$	
+0.22	Hg		$4Br^-$			-2e			$HgBr_4^{-2}$	
+0.22	Mn		F^-			-3e			MnF^{+2}	
+0.23	2As				$3H_2O$	-6e	$6H^+$			As_2O_3
+0.23	Cu		$2Im$			-2e			$Cu(Im)_2^{+2}$	
+0.23	Cu		$2Imdp^{-2}$			-2e			$Cu(Imdp)_2^{-2}$	
+0.23	Fe		F^-			-3e			FeF^{+2}	
+0.23		OH^-	PuO_2OH			-e			$PuO_2(OH)_2$	
+0.23	2Re				$3H_2O$	-6e	$6H^+$			Re_2O_3
+0.23			$(CH_3)_2SO$		H_2O	-2e	$2H^+$			$(CH_3)_2SO_2$
+0.24	Cu		$Adip^{-2}$			-2e			$Cu(Adip)$	

EMF	Elem	Anion	Compound or anion	Cation	H_2O	e =	Cation	Anion	Complex or compound	Compound or element
+0.24	Cu		Suc^{-2}			-2e				Cu(Suc)
+0.24	Cu		$2NH_3$			-2e			$Cu(NH_3)_2^{+2}$	
+0.24	Ge					-2e	Ge^{+2}			
+0.25	As				$2H_2O$	-3e	$3H^+$		$HAsO_2(aq)$	
+0.25	Cu		$3CN^-$			-e			$Cu(CN)_3^{-2}$	
+0.25	Cu		$P_4O_{12}^{-4}$			-2e			$CuP_4O_{12}^{-2}$	
+0.25	Cu		$Tart^{-2}$			-2e			Cu(Tart)	
+0.25	Fe		$2F^-$			-3e			FeF_2^+	
+0.25	Fe		$3Ox^{-2}$			-3e			$Fe(Ox)_3^{-3}$	
+0.25		$2OH^-$	PbO (r)			-2e			PbO_2	H_2O
+0.25	Pt		$4NH_3$			-2e			$Pt(NH_3)_4^{+2}$	
+0.25	Re				$2H_2O$	-4e	$4H^+$		ReO_2	
+0.25				Ru^{+2}		-e	Ru^{+3}			
+0.25	H		$Psfn^-$			-e			Psfn	
+0.26	Cu	Gly^{-2}				-2e				Cu(Gly)
+0.26	Cu		$3Im$			-2e			$Cu(Im)_3^{+2}$	
+0.26			UH_3			-3e	$3H^+$			U
+0.26		$6OH^-$	I^-			-6e		IO_3^-		$3H_2O$
+0.27	2Co	$(FeCy)^{-4}$				-4e			$Co_2(FeCy)$	
+0.27	Cu		Ac^-			-2e			$Cu(Ac)^+$	
+0.27	Cu		$2P_2O_7^{-4}$			-2e			$Cu(P_2O_7)_2^{-6}$	
+0.27	Cu		Pyr			-2e			$Cu(Pyr)^{+2}$	
+0.27	Cu		Fum^{-2}			-2e			Cu(Fum)	
+0.27	Cu		SO_4^{-2}			-2e			$CuSO_4$	
+0.27	Fe		SO_4^{-2}			-3e			$FeSO_4^+$	
+0.27	2Hg		$2Cl^-$			-2e			Hg_2Cl_2	
+0.27	Hg	$4SCN^-$				-2e			$Hg(SCN)_4^{-2}$	
+0.27	Mn		$3Ox^{-2}$			-3e			$Mn(Ox)_3^{-3}$	
+0.27	H		$TMPsfn^-$			-e			TMPsfn	
+0.28	Ag		$S_2O_3^{-2}$			-e			$AgS_2O_3^-$	

EMF	Elem	Anion	Compound or anion	Cation	H_2O	e =	Cation	Anion	Complex or compound	Compound or element
+0.28	Cu	$2Br^-$				-2e				$CuBr_2$
+0.28	Cu		4Im			-2e			$Cu(Im)_4^{+2}$	
+0.28	Cu	Ac^-				-e				CuAc
+0.28	Cu		CHO^-			-2e			$Cu(CHO)^+$	
+0.28	Cu		$4NH_3$			-2e			$Cu(NH_3)_4^{+2}$	
+0.28	Cu		$PhAc^-$			-2e			$Cu(PhAc)^+$	
+0.28	Cu		2Pyr			-2e			$Cu(Pyr)_2^{+2}$	
+0.28	Cu		$HexA^-$			-2e			$Cu(HexA)^+$	
+0.28	Cu		iBu^-			-2e			$Cu(iBu)^+$	
+0.28	Fe		$3F^-$			-3e			FeF_3	
+0.29	Ag	N_3^-				-e				AgN_3
+0.29	H		$Saft^-$			-e				Saft
+0.29	Cu		$BrAc^-$			-2e			$Cu(BrAc)^+$	
+0.29	Cu		$ClAc^-$			-2e			$Cu(ClAc)^+$	
+0.30	3Ag	$(CoCy)^{-3}$				-3e				$Ag_3(CoCy)$
+0.30	Cu		$2P_4O_{12}^{-4}$			-2e			$Cu(P_4O_{12})_2^{-6}$	
+0.30	Cu		3Pyr			-2e			$Cu(Pyr)_3^{+2}$	
+0.30	Cu		$BrBu^-$			-2e			$Cu(BrBu)^+$	
+0.30	H		$InScr^-$			-e				InScr
+0.30	Re					-3e	Re^{+3}			
+0.31	Cu		$2Ac^-$			-2e			$Cu(Ac)_2$	
+0.31	Cu		4Pyr			-2e			$Cu(Pyr)_4^{+2}$	
+0.31	Cu	NO_2^-				-2e			$CuNO_2^+$	
+0.31	Fe		$2SO_4^{-2}$			-3e			$Fe(SO_4)_2^-$	
+0.31	Mn		Cl^-			-3e			$MnCl^{+2}$	
+0.32	Bi				H_2O	-3e	BiO^+			$2H^+$
+0.32			CsH			-e	H^+			Cs
+0.32	Fe		Cl^-			-3e			$FeCl^{+2}$	
+0.32				Mo^{+3}	$3H_2O$	-3e	$6H^+$		MoO_3	

EMF	Elem	Anion	Compound or anion	Cation	H_2O	e =	Cation	Anion	Complex or compound	Compound or element
+0.32			MoO_2		H_2O	-2e	$2H^+$			MoO_3
+0.32			RbH			-e	H^+			Rb
+0.33		$2OH^-$	ClO_2^-			-2e		ClO_3^-		H_2O
+0.33	Cu		Br^-			-2e			$CuBr^+$	
+0.33	Cu		Nac^-			-2e			$Cu(Nac)^+$	
+0.33	Cu		SO_4^{-2}			-2e			$CuSO_4$	
+0.33	3Cu	$2PO_4^{-3}$				-6e				$Cu_3(PO_4)_2$
+0.33	Fe		$2Cl^-$			-3e			$FeCl_2^+$	
+0.33				U^{+4}	$2H_2O$	-2e	$4H^+$		UO_2^{+2}	
+0.33	H		$NuRd^-$			-e			NuRd	
+0.34	Ag		Tate			-e			$Ag(Tate)^+$	
+0.34	3Ag	PO_4^{-3}				-3e				Ag_3PO_4
+0.34	Ag		Teta			-e			$Ag(Teta)^+$	
+0.34	Cu					-2e	Cu^{+2}			
+0.34	Cu		Cl^-			-2e			$CuCl^+$	
+0.34	Fe		Br^-			-3e			$FeBr^{+2}$	
+0.35	Ag		IO_3^-			-e			$AgIO_3$	
+0.35	2Ag	$2OH^-$				-2e			Ag_2O	H_2U
+0.35	Cu		$5NH_3$			-2e			$Cu(NH_3)_5^{+2}$	
+0.36		$2OH^-$	ClO_3^-			-2e		ClO_4^-		H_2O
+0.36	Cu		$2Cl^-$			-2e			$CuCl_2$	
+0.36	Fe		$3Cl^-$			-3e			$FeCl_3$	
+0.36			$(FeCy)^{-4}$			-e		$(FeCy)^{-3}$		
+0.36	Re				$4H_2O$	-7e	$8H^+$	ReO_4^-		
+0.36	S				$4H_2O$	-6e	$8H^+$	SO_4^{-2}		
+0.36				V^{+3}	H_2O	-e	$2H^+$		VO_2^+	
+0.37	2Bi				$3H_2O$	-6e	$6H^+$			Bi_2O_3
+0.37			HCN (aq)			-e	H^+		$\frac{1}{2}C_2N_2(g)$	
+0.38	Po	$4Cl^-$				-2e		$PoCl_4^{-2}$		
+0.39				$2NH_3OH^+$		-4e	$6H^+$		$H_2N_2O_2$	
+0.39	2Hg	$2IO_3^-$				-2e				$Hg_2(IO_3)_2$

EMF	Elem	Anion	Compound or anion	Cation	H_2O	e =	Cation	Anion	Complex or compound	Compound or element
+0.39	H		$RsSf6^-$			-e			$RsSf6$	
+0.39			KH			-e	H^+			K
+0.39			NaH			-e	H^+			Na
+0.40			$4OH^-$			-4e			$2H_2O$	O_2
+0.40	Au	$2OH^-$				-e		$Au(OH)_2^-$		
+0.40	Cu	$2S_2O_3^{-2}$				-e		$Cu(S_2O_3)_2^{-3}$		
+0.40			$S_2O_3^{-2}$		$3H_2O$	-4e	$2H^+$		$2H_2SO_3$	
+0.40	Pt	$4I^-$				-2e		PtI_4^-		
+0.40	Tc					-2e	Tc^{+2}			
+0.40		$2OH^-$	TeO_3^{-2}			-2e		TeO_4^{-2}		H_2O
+0.40	H		$Chxo^-$			-e			$Chxo$	
+0.41	H		Cl^-			-e			HCl	
+0.41		OH^-	HO_2^-			-e		O_2^-	H_2O	
+0.42	Cu		$4CN^-$			-e			$Cu(CN)_4^{-3}$	
+0.42	Cu		$3S_2O_3^{-2}$			-e			$Cu(S_2O_3)_3^{-5}$	
+0.43	Rh		$6Cl^-$			-3e		$RhCl_6^{-3}$		
+0.43	H		$Acto^-$			-e			$Acto$	
+0.44	Ag		$Deta$			-e			$Ag(Deta)^+$	
+0.45	Cu		$3Pyr$			-e			$Cu(Pyr)_3^+$	
+0.45	S				$3H_2O$	-4e	$4H^+$		H_2SO_3	
+0.46	2Ag		CrO_4^{-2}			-2e			Ag_2CrO_4	
+0.46	2Ag	MoO_4^{-2}				-2e				Ag_2MoO_4
+0.46	Cu		$4Pyr$			-e			$Cu(Pyr)_4^+$	
+0.46	Cu		$2SO_3^{-2}$			-e			$Cu(SO_3)_2^{-3}$	
+0.46	Ru					-2e	Ru^{+2}			
+0.46		$W(CN)_8^{-4}$				-e		$W(CN)_8^{-3}$		
+0.47	2Ag		CO_3^{-2}			-2e			Ag_2CO_3	
+0.47	2Ag	WO_4^{-2}				-2e				Ag_2WO_4
+0.47	Ag		Ptn			-e			$Ag(Ptn)^+$	
+0.47	H		ClO_4^-			-e			$HClO_4$	
+0.47	2S				$3H_2O$	-4e	$6H^+$	$S_2O_3^{-2}$		
+0.48	Cu		$3SO_3^{-2}$			-e			$Cu(SO_3)_3^{-5}$	

EMF	Elem	Anion	Compound or anion	Cation	H_2O	e =	Cation	Anion	Complex or compound	Compound or element
+0.48	C				$3H_2O$	-4e	$6H^+$	CO_3^{-2}		
+0.48			Sb_2O_4 (c)		H_2O	-2e	$2H^+$		Sb_2O_5(c)	
+0.49	2Ag		MoO_4^{-2}			-2e			Ag_2MoO_4	
+0.49	Ag		SO_3^{-2}			-e			$AgSO_3^-$	
+0.49		$2OH^-$	I^-			-2e		IO^-	H_2O	
+0.49		$2OH^-$	$Ni(OH)_2$			-2e		NiO_2		$2H_2O$
+0.50	Ag		En			-e			$Ag(En)^+$	
+0.50		AtO^-	$4OH^-$			-4e		AtO_3^-	$2H_2O$	
+0.51			ReO_2		$2H_2O$	-3e	$4H^+$	ReO_4^-		
+0.51	2Hg	$2Ac^-$				-2e				$Hg_2(Ac)_2$
+0.51			$S_4O_6^{-2}$		$6H_2O$	-6e	$4H^+$	$4H_2SO_3$		
+0.52			C_2H_6 (g)			-2e	$2H^+$	C_2H_4 (g)		
+0.52	C				H_2O	-2e	$2H^+$			CO
+0.52	Cu					-e	Cu^+			
+0.53	H		Br^-			-e			HBr	
+0.53	Te				$2H_2O$	-4e	$4H^+$		TeO_2 (c)	
+0.54			CuCl			-e	Cu^{+2}	Cl^-		
+0.54			$2I^-$			-2e				I_2 (c)
+0.54			$3I^-$			-2e		I_3^-		
+0.55	Ag		Br^-			-e			AgBr	
+0.55	Ag		BrO_3^-			-e			$AgBrO_3$	
+0.55	Pd	$5Cl^-$				-2e		$PdCl_5^{-3}$		
+0.56	Ag		NO_2^-			-e			$AgNO_2$	
+0.56			$HAsO_2$		$2H_2O$	-2e	$2H^+$		H_3AsO_4	
+0.56	Au	$4I^-$				-3e		AuI_4^-		
+0.56			MnO_2^{-2}			-e		MnO_4^-		
+0.56	Po					-3e	Po^{+3}			
+0.56	Te				$2H_2O$	-4e	$3H^+$		$TeOOH^+$	
+0.57	Ag		$2NH_3$			-e			$Ag(NH_3)_2^+$	
+0.57	Pd	$3Cl^-$				-2e		$PdCl_3^-$		
+0.57			$2H_2SO_3$			-2e	$4H^+$	$S_2O_6^{-2}$		

EMF	Elem	Anion	Compound or anion	Cation	H_2O	e =	Cation	Anion	Complex or compound	Compound or element
+0.57	Te					-4e	Te^{+4}			
+0.58	Ag		$Alan^-$			-e			$Ag(Alan)$	
+0.58	Au	$2I^-$				-e		AuI_2^-		
+0.58	Pt	$4Br^-$				-2e		$PtBr_4^{-2}$		
+0.58				$2SbO^+$	$3H_2O$	-4e	$6H^+$		$Sb_2O_5(c)$	
+0.59	Ag		$2Alan^-$			-e			$Ag(Alan)_2^-$	
+0.59	Ag		Gl^-			-e			$Ag(Gl)$	
+0.59	As	$3Br^-$				-3e					$AsBr_3$
+0.59			CH_4 (g)		H_2O	-2e	$2H^+$		$CH_3OH(aq)$	
+0.59		$4OH^-$	MnO_2			-3e		MnO_4^-	$2H_2O$	
+0.60	Ag		$2Gl^-$			-e			$Ag(Gl)_2^-$	
+0.60		$4OH^-$	MnO_2			-2e		MnO_4^{-2}	$2H_2O$	
+0.60	Pd	$4Br^-$				-2e		$PdBr_4^{-2}$		
+0.60	Rh					-e	Rh^+			
+0.60				Rh^+		-e	Rh^{+2}			
+0.60	Ru	$5Cl^-$				-3e		$RuCl_5^{-2}$		
+0.60	Rh					-2e	Rh^{+2}			
+0.60			RuO_4^{-2}			-e		RuO_4^-		
+0.60				Tc^{+2}	$2H_2O$	-2e	$4H^+$		TcO_2	
+0.61	Ag		NH_3			-e			$Ag(NH_3)^+$	
+0.61		$2OH^-$	Ag_2O			-2e			$2AgO$.	H_2O
+0.61		$6OH^-$	Br^-			-6e		BrO_3^-	$3H_2O$	
+0.61	Pt	$3Cl^-$	NH_3			-2e		$Pt(NH_3)Cl_3^-$		
+0.62	Ag		$2Br^-$			-e			$AgBr_2^-$	
+0.62				Cu^+	H_2O	-e	$2H^+$			CuO
+0.62	2Hg		SO_4^{-2}			-2e			Hg_2SO_4	
+0.62	Pd		$4Cl^-$			-2e		$PdCl_4^{-2}$		
+0.62				U^{+4}	$2H_2O$	-e	$4H^+$		UO_2^+	
+0.63	Ag		$2En$			-e			$Ag(En)_2^+$	
+0.63	C				$2H_2O$	-2e	$2H^+$			$HCOOH$
+0.63	Tl		OH^-			-3e			$TlOH^{+2}$	

EMF	Elem	Anion	Compound or anion	Cation	H_2O	e =	Cation	Anion	Complex or compound	Compound or element
+0.64	Ag		Cl^-			-e			AgCl	
+0.64	Ag		$Glgl^-$			-e			Ag(Glgl)	
+0.64	2Hg	HPO_4^{-2}				-2e				Hg_2HPO_4
+0.64			CuBr			-e	Cu^{+2}	Br^-		
+0.64	Pd	$2Cl^-$				-2e				$PdCl_2$
+0.65	Ag		$3CNS^-$			-e			$Ag(CNS)_3^{-2}$	
+0.65	2Ag		SO_4^{-2}			-2e			Ag_2SO_4	
+0.65	H		I^-			-e			HI	
+0.65	Po					-2e	Po^{+2}			
+0.65	Tl		$2OH^-$			-3e			$Tl(OH)_2^+$	
+0.66	Ag		OH^-			-e			AgOH	
+0.66	Au		$4CNS^-$			-3e			$Au(CNS)_4^-$	
+0.66	Pt	$5Cl^-$	NH_3			-4e			$Pt(NH_3)Cl_5^-$	
+0.66		$2OH^-$	ClO^-			-2e	ClO_2^-	H_2O		
+0.67	Ag		$2Glgl^-$			-e			$Ag(Glgl)_2^-$	
+0.67			Cu_2O		H_2O	-2e	$2H^+$			2CuO(a)
+0.68	Ag		$2Cl^-$			-e			$AgCl_2^-$	
+0.68	Ag		Pyr			-e			$Ag(Pyr)^+$	
+0.68	Ag		2Pyr			-e			$Ag(Pyr)_2^+$	
+0.68	Ag		$2SO_3^{-2}$			-e			$Ag(SO_3)_2^{-3}$	
+0.68			H_2O_2(aq)			-2e	$2H^+$			O_2 (g)
+0.68		$2Cl^-$	$PtCl_4^{-2}$			-2e		$PtCl_6^{-2}$		
+0.68	Ru				$4H_2O$	-4e	$4H^+$			$Ru(OH)_4$
+0.69	Ag		$3Br^-$			-e			$AgBr_3^{-2}$	
+0.69	Os				$2H_2O$	-4e	$4H^+$			OsO_2
+0.70	At_2				$2H_2O$	-2e	$2H^+$			2HAtO
+0.70		$3OH^-$	IO_3^-			-2e		$H_3IO_6^{-2}$		
+0.70				$3NH_4^+$		-8e	$11H^+$		HN_3 (aq)	
+0.70			TcO_2		$2H_2O$	-3e	$4H^+$	TcO_4^-		
+0.70			$C_6H_4(OH)_2$			-2e	$2H^+$		$C_6H_4O_2$	
+0.71			$H_2N_2O_2$			-2e	$2H^+$		2NO	
+0.71			OH (g)		H_2O	-e	H^+		H_2O_2(aq)	

EMF	Elem	Anion	Compound or anion	Cation	H_2O	e =	Cation	Anion	Complex or compound	Compound or element
+0.72		$5OH^-$	$Fe(OH)_3$			-3e		FeO_4^{-2}	$4H_2O$	
+0.72			SrH_2			-2e	$2H^+$			Sr
+0.72	Tl		Br^-			-3e			$TlBr^{+2}$	
+0.73	Ag		$3SeCN^-$			-e			$Ag(SeCN)_3^{-2}$	
+0.73			C_2H_4 (g)			-2e	$2H^+$		C_2H_2 (g)	
+0.73			LiH			-e	H^+			Li
+0.73		$2OH^-$	N_2H_4			-2e			$2NH_2OH$	
+0.73	Pt		$4Cl^-$			-2e		$PtCl_4^{-2}$		
+0.74	Ag		$3CN^-$			-e			$Ag(CN)_3^{-2}$	
+0.74	Ag		$4CNS^-$			-e			$Ag(CNS)_4^{-3}$	
+0.74		$2OH^-$	$2AgO$			-2e			Ag_2O	H_2O
+0.74	2Ru				$3H_2O$	-6e	$6H^+$			Ru_2O_3
+0.74	Se				$3H_2O$	-4e	$4H^+$	$H_2SeO_3(aq)$		
+0.75				Np^{+4}	$2H_2O$	-e	NpO_2^+	$4H^+$		
+0.75			Cu_2O		H_2O	-2e	$2H^+$			2CuO(b)
+0.75	Tl		Cl^-			-3e			$TlCl^{+2}$	
+0.76	Ag		Ac^-			-e			Ag(Ac)	
+0.76		$2OH^-$	Br^-			-2e		BrO^-	H_2O	
+0.76	Po					-4e	Po^{+4}			
+0.77			$2CNS^-$			-2e			$(CNS)_2$	
+0.77				Fe^{+2}		-e	Fe^{+3}			
+0.77	Ir		$6Cl^-$			-3e		$IrCl_6^{-3}$		
+0.77	Pd	Cl^-				-2e			$PdCl^+$	
+0.77	Tl		$2Br^-$			-3e			$TlBr_2^+$	
+0.78	Ag	F^-				-e				AgF
+0.78	Ag		$4Br^-$			-e			$AgBr_4^{-3}$	
+0.78			CaH_2			-2e	$2H^+$			Ca
+0.79	Ag		$2Ac^-$			-e			$Ag(Ac)_2^-$	
+0.79	Ag	ClO_4^-				-e				$AgClO_4$
+0.79	Ag		SO_4^{-2}			-e			$AgSO_4^-$	
+0.79	2Hg					-2e	Hg^{+2}			
+0.80	Ag					-e	Ag^+			

EMF	Elem	Anion	Compound or anion	Cation	H_2O	e =	Cation	Anion	Complex or compound	Compound or element
+0.80	Ag		$4Cl^-$			-e			$AgCl_4^{-3}$	
+0.80	Ag	$2SO_4^{-2}$				-e			$Ag(SO_4)_2^{-3}$	
+0.80			N_2O_4 (g)		$2H_2O$	-2e	$4H^+$	$2NO_3^-$		
+0.80				Po^{+2}	$2H_2O$	-2e	$4H^+$			PoO_2
+0.80	Rh					-3e	Rh^{+3}			
+0.80	2Rh				H_2O	-2e	$2H^+$				Rh_2O
+0.80	Tl		$2Cl^-$			-3e			$TlCl_2^+$	
+0.81	Rh				H_2O	-2e	$2H^+$				RhO
+0.82	Tl		$3Br^-$			-3e			$TlBr_3$	
+0.83	Ag		$4CN^-$			-e			$Ag(CN)_4^{-3}$	
+0.85	Os				$4H_2O$	-8e	$8H^+$			OsO_4	
+0.85	Os				$5H_2O$	-8e	$8H^+$				H_2OsO_5
+0.86			CuI			-e	Cu^{+2}	I^-		
+0.86			$H_2N_2O_2$		$2H_2O$	-4e	$4H^+$			$2HNO_2$	
+0.86		$RuCy^{-4}$				-e		$RuCy^{-3}$		
+0.86	Tl		$4Br^-$			-3e			$TlBr_4^-$	
+0.86	Tl		$4Cl^-$			-3e			$TlCl_4^-$	
+0.87	Au		$4Br^-$			-3e		$AuBr_4^-$		
+0.87	Ir				H_2O	-2e	$2H^+$				IrO
+0.87	2Rh				$3H_2O$	-6e	$6H^+$			Rh_2O_3	
+0.87	Tl		$3Cl^-$			-3e			$TlCl_3$	
+0.88			$3OH^-$			-2e		HO_2^-	H_2O	
+0.88	Hg	F^-				-2e			HgF^+	
+0.89		$2OH^-$	Cl^-			-2e		ClO^-	H_2O	
+0.90		$4OH^-$	$HXeO_4^-$			-2e		$HXeO_6^{-3}$	$2H_2O$	
+0.90	Xe	$7OH^-$				-6e		$HXeO_4^-$	$3H_2O$	
+0.91	Tl	NO_3^-				-3e			$TlNO_3^{+2}$	
+0.92				Hg_2^{+2}		-2e	$2Hg^{+2}$			
+0.93	Hg				H_2O	-2e	$2H^+$				HgO
+0.93				PuO_2^+		-e	PuO_2^{+2}			
+0.93	Ir				$2H_2O$	-4e	$4H^+$				IrO_2
+0.93	2Ir				$3H_2O$	-3e	$6H^+$				Ir_2O_3

EMF	Elem	Anion	Compound or anion	Cation	H₂O	e =	Cation	Anion	Complex or compound	Compound or element
+0.94			HNO_2		H_2O	-2e	$3H^+$	NO_3^-		
+0.96	Au		$2Br^-$			-e		$AuBr_2^-$		
+0.96			NO		$2H_2O$	-3e	$4H^+$	NO_3^-		
+0.97				Pu^{+3}		-e	Pu^{+4}			
+0.98	Pt				$2H_2O$	-2e	$2H^+$		$Pt(OH)_2$	
+0.98	Pt				H_2O	-2e	$2H^+$			PtO
+0.99			$IrBr_6^{-4}$			-e		$IrBr_6^{-3}$		
+0.99			2HCl			-2e	$2H^+$			Cl_2
+0.99	Pd					-2e	Pd^{+2}			
+1.00	Au		$4Cl^-$			-3e		$AuCl_4^-$		
+1.00			NO		H_2O	-e	H^+		HNO_2	
+1.00				VO^{+2}	$3H_2O$	-e	$V(OH)_4^+$		$2H^+$	
+1.02			$IrCl_6^{-3}$			-e		$IrCl_6^{-2}$		
+1.02			TeO_2		$4H_2O$	-2e	$2H^+$		$H_6TeO_6(c)$	
+1.03	Hg				$2H_2O$	-2e	$2H^+$			$Hg(OH)_2$
+1.03			2NO		$2H_2O$	-4e	$4H^+$	N_2O_4		
+1.04				Pu^{+4}	$2H_2O$	-2e	PuO_2^{+2}		$4H^+$	
+1.06	½I_2		$2Cl^-$			-e		ICl_2^-		
+1.07			$2Br^-$			-2e				Br_2 (1)
+1.07			HNO_2			-2e	$2H^+$		N_2O_4	
+1.09			$2Br^-$			-2e				Br_2(aq)
+1.09			HNO_2			-e	H^+			NO_2
+1.10			$Cu(CN)_2^-$			-e	Cu^{+2}	$2CN^-$		
+1.10			$Pt(OH)_2$			-2e	$2H^+$		PtO_2	
+1.10	Rh				$4H_2O$	-6e	$8H^+$	RhO_4^{-2}		
+1.13	Cu				$2H_2O$	-2e	$3H^+$	$HCuO_2^-$		
+1.15			H_2SeO_3		H_2O	-2e	$4H^+$		SeO_4^{-2}	
+1.15				NpO_2^+		-e	NpO_2^{+2}			
+1.15	Au	$2Cl^-$				-e		$AuCl_2^-$		
+1.15				Pu^{+4}	$2H_2O$	-e	PuO_2^+		$4H^+$	
+1.16			ClO_2^-			-e			ClO_2 (g)	
+1.16	Ir					-3e	Ir^{+3}			

EMF	Elem	Anion	Compound or anion	Cation	H_2O	e =	Cation	Anion	Complex or compound	Compound or element
+1.18	C		$4Cl^-$	$4H^+$		-4e	$4H^+$		CCl_4	
+1.19			ClO_3^-		H_2O	-2e	$2H^+$	ClO_4^-		
+1.19			$2H_2O$ (g)			-4e	$4H^+$			O_2
+1.20				Rh^{+2}		-e	Rh^{+3}			
+1.20	$\frac{1}{2}I_2$				$3H_2O$	-5e	$6H^+$	IO_3^-		
+1.20		$(RhCl_6)^{-3}$				-e		$(RhCl_6)^{-2}$		
+1.20	Pt					-2e	Pt^{+2}			
+1.21			$HClO_2$		H_2O	-2e	$3H^+$	ClO_3^-		
+1.23			$2H_2O$ (1)			-4e	$4H^+$			O_2
+1.23				Mn^{+2}	$2H_2O$	-2e	$4H^+$		MnO_2	
+1.23	2S		$2Cl^-$			-2e			S_2Cl_2	
+1.24	O_2		$2OH^-$			-2e			H_2O	O_3 (g)
+1.25				Tl^+		-2e	Tl^{+3}			
+1.26				Am^{+4}	$2H_2O$	-e	AmO_2^+		$4H^+$	
+1.26	I_2					-2e	$2I^+$			
+1.28			$HClO_2$			-e	H^+		ClO_2	
+1.28				$2NH_4^+$		-2e	$N_2H_5^+$		$3H^+$	
+1.29			N_2O (g)		$3H_2O$	-4e	$4H^+$		$2HNO_2$(aq)	
+1.29		$2Cl^-$	$PdCl_4^{-2}$			-2e		$PdCl_6^{-2}$		
+1.33				$2Cr^{+3}$	$7H_2O$	-6e	$14H^+$	$Cr_2O_7^{-2}$		
+1.35				NH_4^+	H_2O	-2e	NH_3OH^+		$2H^+$	
+1.36	2Au				$3H_2O$	-6e	$6H^+$			Au_2O_3
+1.36			$2Cl^-$			-2e				Cl_2
+1.40			HAtO		$2H_2O$	-4e	$4H^+$		$HAtO_3$	
+1.40				Au^+		-2e	Au^{+3}			
+1.42				$N_2H_5^+$	$2H_2O$	-2e	H^+		$2NH_3OH^+$	
+1.44	F_2				H_2O	-2e	$2H^+$			F_2O
+1.45	Au				$3H_2O$	-3e	$3H^+$		$Au(OH)_3$(c)	
+1.45	Au				$3H_2O$	-3e	$3H^+$			$Au(OH)_3$
+1.45	$\frac{1}{2}I_2$				H_2O	-e	H^+		HIO	
+1.45	N_2				$4H_2O$	-6e	$6H^+$			$2HNO_2$
+1.46				Pb^{+2}	$2H_2O$	-2e	$4H^+$		PbO_2	

EMF	Elem	Anion	Compound or anion	Cation	H_2O	e =	Cation	Anion	Complex or compound	Compound or element
+1.50	Au					-3e	Au^{+3}			
+1.50			$H_2O_2(aq)$			-e	H^+		HO_2 (aq)	
+1.51				Mn^{+2}		-e	Mn^{+3}			
+1.51				Mn^{+2}	$4H_2O$	-5e	$8H^+$	MnO_4^-		
+1.52	½Br$_2$				$3H_2O$	-5e	$6H^+$	BrO_3^-		
+1.52	Cu				$2H_2O$	-2e	$4H^+$	CuO_2^{-2}		
+1.52			PoO_2		H_2O	-2e	$2H^+$			PoO_3
+1.59				$2BiO^+$	$2H_2O$	-2e	$4H^+$			Bi_2O_4
+1.60				Bk^{+3}		-e	Bk^{+4}			
+1.60	½Br$_2$				H_2O	-e	H^+			HBrO
+1.60			IO_3^-		$3H_2O$	-2e	H^+			H_5IO_6
+1.61				Ce^{+3}		-e	Ce^{+4}			
+1.63	½Cl$_2$				H_2O	-e	H^+			HClO
+1.64				AmO_2^+		-e	AmO_2^{+2}			
+1.65			HClO		H_2O	-2e	$2H^+$			$HClO_2$
+1.67				Pb^{+2}		-2e	Pb^{+4}			
+1.68	Ni				$2H_2O$	-2e	$4H^+$			NiO_2
+1.68			$PbSO_4$		$2H_2O$	-2e	$4H^+$	SO_4^{-2}	PbO_2	
+1.69				Am^{+3}	$2H_2O$	-3e	AmO_2^{+2}		$4H^+$	
+1.69	Au					-e	Au^+			
+1.70			MnO_2		$2H_2O$	-3e	$4H^+$	MnO_4^-		
+1.71	Cl$_2$				H_2O	-2e	$2H^+$			Cl_2O
+1.72				Am^{+3}	$2H_2O$	-2e	AmO_2^+		$4H^+$	
+1.73				Cu^+	$2H_2O$	-e	$3H^+$	$HCuO_2^-$		
+1.77				Ag^+	H_2O	-e	$2H^+$			AgO
+1.78					$2H_2O$	-2e	$2H^+$		H_2O_2	
+1.78			Cu_2O		$3H_2O$	-2e	$4H^+$	$2HCuO_2^-$		
+1.80	Xe				$3H_2O$	-6e	$6H^+$		XeO_3	
+1.81				Co^{+2}		-e	Co^{+3}			
+1.96	N$_2$			NH_4^+		-2e	$3H^+$		HN_3	
+1.97	3Pb	$2PO_4^{-3}$				-6e				$Pb_3(PO_4)_2$
+1.98				Ag^+		-e	Ag^{+2}			

EMF	Elem	Anion	Compound or anion	Cation	H_2O	e =	Cation	Anion	Complex or compound	Compound or element
+2.00				Ag^+	H_2O	-2e	$2H^+$			AgO^+
+2.01			$2SO_4^{-2}$			-2e		$S_2O_8^{-2}$		
+2.02			OH^-			-e			OH (g)	
+2.02				Ag^{+2}	H_2O	-e	$2H^+$			AgO^+
+2.07	O_2				H_2O	-2e	$2H^+$			O_3
+2.15			$2F^-$		H_2O	-4e	$2H^+$		F_2O	
+2.18				Am^{+3}		-e	Am^{+4}			
+2.20				Fe^{+3}	$4H_2O$	-3e	$8H^+$	FeO_4^{-2}		
+2.22	Ag				H_2O	-e	$2H^+$	AgO^-		
+2.42					H_2O	-2e	$2H^+$			O (g)
+2.51				Cu^+	$2H_2O$	-e	$4H^+$	CuO_2^{-2}		
+2.56			Cu_2O		$3H_2O$	-2e	$6H^+$	$2CuO_2^{-2}$		
+2.65	N_2				$2H_2O$	-2e	$2H^+$		$H_2N_2O_2$	
+2.76				Pr^{+3}	$2H_2O$	-e	$4H^+$			PrO_2
+2.85					H_2O	-e	H^+		OH	
+2.86				Pr^{+3}		-e	Pr^{+4}			
+2.87			$2F^-$			-2e				F_2 (g)
+3.00			XeO_3		$3H_2O$	-2e	$2H^+$		H_4XeO_6	
+3.06			2HF (aq)			-2e	$2H^+$			F_2 (g)

PART II

CHEMICAL ELECTRODE

POTENTIALS BY ELEMENT

(Arranged Alphabetically By Symbol)

EMF	Elem	Anion	Compound or anion	Cation	H_2O	e =	Cation	Anion	Complex or compound	Compound or element
-2.60	Ac					-3e	Ac^{+3}			

EMF	Elem	Anion	Compound or anion	Cation	H_2O	e =	Cation	Anion	Complex or compound	Compound or element
-0.66	2Ag		S^{-2}			-2e			Ag_2S	
-0.31	Ag		$2CN^-$			-e			$Ag(CN)_2^-$	
-0.15	Ag		I^-			-e			AgI	
-0.02	Ag		CN^-			-e			$AgCN$	
+0.02	Ag		$2S_2O_3^{-2}$			-e			$Ag(S_2O_3)_2^{-3}$	
+0.07	Ag		Br^-			-e			$AgBr$	
+0.15	4Ag		$(FeCy)^{-4}$			-4e			$Ag_4(FeCy)$	
+0.28	Ag		$S_2O_3^{-2}$			-e			$AgS_2O_3^-$	
+0.29	Ag		N_3^-			-e			AgN_3	
+0.30	3Ag		$(CoCy)^{-3}$			-3e			$Ag_3(CoCy)$	
+0.34	Ag		Tate			-e			$Ag(Tate)^+$	
+0.34	3Ag		PO_4^{-3}			-3e			Ag_3PO_4	
+0.34	Ag		Teta			-e			$Ag(Teta)^+$	
+0.35	Ag		IO_3^-			-e			$AgIO_3$	
+0.35	2Ag		$2OH^-$			-2e			Ag_2O	H_2O
+0.44	Ag		Deta			-e			$Ag(Deta)^+$	
+0.46	2Ag		CrO_4^{-2}			-2e			Ag_2CrO_4	
+0.46	2Ag		MoO_4^{-2}			-2e			Ag_2MoO_4	
+0.47	2Ag		CO_3^{-2}			-2e			Ag_2CO_3	
+0.47	Ag		Ptn			-e			$Ag(Ptn)^+$	
+0.47	2Ag		WO_4^{-2}			-2e			$AgWO_4$	
+0.49	2Ag		MoO_4^{-2}			-2e			Ag_2MoO_4	
+0.49	Ag		SO_3^{-2}			-e			$AgSO_3^-$	
+0.50	Ag		En			-e			$Ag(En)^+$	
+0.55	Ag		Br^-			-e			$AgBr$	
+0.55	Ag		BrO_3^-			-e			$AgBrO_3$	
+0.56	Ag		NO_2^-			-e			$AgNO_2$	
+0.57	Ag		$2NH_3$			-e			$Ag(NH_3)_2^+$	

EMF	Elem	Anion	Compound or anion	Cation	H_2O	e =	Cation	Anion	Complex or compound	Compound or element
+0.58	Ag		$Alan^-$			-e			$Ag(Alan)$	
+0.59	Ag		$2Alan^-$			-e			$Ag(Alan)_2^-$	
+0.59	Ag		Gl^-			-e			$Ag(Gl)$	
+0.60	Ag		$2Gl^-$			-e			$Ag(Gl)_2^-$	
+0.61	Ag		NH_3			-e			$Ag(NH_3)^+$	
+0.61		$2OH^-$	Ag_2O			-2e			$2AgO$	H_2O
+0.62	Ag		$2Br^-$			-e			$AgBr_2^-$	
+0.63	Ag		$2En$			-e			$Ag(En)_2^+$	
+0.64	Ag		Cl^-			-e			$AgCl$	
+0.64	Ag		$Glgl^-$			-e			$Ag(Glgl)$	
+0.65	Ag		$3CNS^-$			-e			$Ag(CNS)_3^{-2}$	
+0.65	2Ag		SO_4^{-2}			-2e			Ag_2SO_4	
+0.66	Ag		OH^-			-e			$AgOH$	
+0.67	Ag		$2Glgl^-$			-e			$Ag(Glgl)_2^-$	
+0.68	Ag		$2Cl^-$			-e			$AgCl_2^-$	
+0.68	Ag		Pyr			-e			$Ag(Pyr)^+$	
+0.68	Ag		$2Pyr$			-e			$Ag(Pyr)_2^+$	
+0.68	Ag		$2SO_3^{-2}$			-e			$Ag(SO_3)_2^{-3}$	
+0.69	Ag		$3Br^-$			-e			$AgBr_3^{-2}$	
+0.73	Ag		$3SeCN^-$			-e			$Ag(SeCN)_3^{-2}$	
+0.74	Ag		$3CN^-$			-e			$Ag(CN)_3^{-2}$	
+0.74	Ag		$4CNS^-$			-e			$Ag(CNS)_4^{-3}$	
+0.74		$2OH^-$	$2AgO$			-2e			Ag_2O_3	H_2O
+0.76	Ag		Ac^-			-e			$Ag(Ac)$	
+0.78	Ag		F^-			-e			AgF	
+0.78	Ag		$4Br^-$			-e			$AgBr_4^{-3}$	
+0.79	Ag		$2Ac^-$			-e			$Ag(Ac)_2^-$	
+0.79	Ag		ClO_4^-			-e			$AgClO_4$	
+0.79	Ag		SO_4^{-2}			-e			$AgSO_4^-$	
+0.80	Ag					-e	Ag^+			
+0.80	Ag		$4Cl^-$			-e			$AgCl_4^{-3}$	

EMF	Elem	Anion	Compound or anion	Cation	H_2O	e =	Cation	Anion	Complex or compound	Compound or element
+0.80	Ag		$2SO_4^{-2}$			-e			$Ag(SO_4)_2^{-3}$	
+0.83	Ag		$4CN^-$			-e			$Ag(CN)_4^{-3}$	
+1.77				Ag^+	H_2O	-e	$2H^+$		AgO	
+1.98				Ag^+		-e	Ag^{+2}			
+2.00				Ag^+	H_2O	-2e	AgO^+			$2H^+$
+2.02				Ag^{+2}	H_2O	-e	AgO^+			$2H^+$
+2.22	Ag				H_2O	-e	$2H^+$	AgO^-		

EMF	Elem	Anion	Compound or anion	Cation	H_2O	e =	Cation	Anion	Complex or compound	Compound or element
-2.33	Al		$4OH^-$			-3e		$H_2AlO_3^-$		H_2O
-2.30	Al		$3OH^-$			-3e			$Al(OH)_3$	
-2.07	Al		$6F^-$			-3e			AlF_6^{-3}	
-2.02	Al		$Data^{-4}$			-3e			$Al(Data)^-$	
-1.95	Al		Sal^{-2}			-3e			$Al(Sal)^+$	
-1.84	Al		$AcAc^-$			-3e			$Al(AcAc)^{+2}$	
-1.84	Al		OH^-			-3e			$AlOH^{+2}$	
-1.83	Al		$2AcAc^-$			-3e			$Al(AcAc)_2^+$	
-1.80	Al		Cl^-			-3e			$AlCl^{+2}$	
-1.79	Al		F^-			-3e			AlF^{+2}	
-1.78	Al		$3AcAc^-$			-3e			$Al(AcAc)_3$	
-1.77	Al		$2F^-$			-3e			AlF_2^+	
-1.75	Al		$3F^-$			-3e			AlF_3	
-1.74	Al		$3Ox^{-2}$			-3e			$Al(Ox)_3^{-3}$	
-1.72	Al		$4F^-$			-3e			AlF_4^-	
-1.70	Al		$5F^-$			-3e			AlF_5^{-2}	
-1.68	Al		$6F^-$			-3e			AlF_6^{-3}	
-1.68	Al		Nac^-			-3e			$Al(Nac)^{+2}$	
-1.66	Al					-3e	Al^{+3}			
-1.54	2Al				$6H_2O$	-6e	$6H^+$		Al_2O_3	$3H_2O$
-1.26	Al				$2H_2O$	-3e	$4H^+$	AlO_2^-		

EMF	Elem	Anion	Compound or anion	Cation	H_2O	e =	Cation	Anion	Complex or compound	Compound or element
-2.32	Am					-3e	Am^{+3}			
-1.88	Am				$3H_2O$	-3e	$3H^+$		$Am(OH)_3$	
-1.68	2Am				$3H_2O$	-6e	$6H^+$		Am_2O_3	
+1.26				Am^{+4}	$2H_2O$	-e	AmO_2^+			$4H^+$
+1.64				AmO_2^+		-e	AmO_2^{+2}			
+1.69				Am^{+3}	$2H_2O$	-3e	AmO_2^{+2}			$4H^+$
+1.72				Am^{+3}	$2H_2O$	-2e	AmO_2^+			$4H^+$
+2.18				Am^{+3}		-e	Am^{+4}			

EMF	Elem	Anion	Compound or anion	Cation	H_2O	e =	Cation	Anion	Complex or compound	Compound or element
-0.68	As		$4OH^-$			-3e		AsO_2^-	$2H_2O$	
-0.68		AsO_2^-	$4OH^-$			-2e		AsO_4^{-3}	$2H_2O$	
-0.61			AsH_3 (g)			-3e	$3H^+$			As
-0.57	H		AsO_2^-			-e			$HAsO_2$	
-0.14	H		AsO_3^-			-e			$HAsO_3$	
+0.23	2As				$3H_2O$	-6e	$6H^+$		As_2O_3	
+0.25	As				$2H_2O$	-3e	$3H^+$		$HAsO_2$ (aq)	
+0.56			$HAsO_2$		$2H_2O$	-2e	$2H^+$		H_3AsO_4	
+0.59	As		$3Br^-$			-3e			$AsBr_3$	

EMF	Elem	Anion	Compound or anion	Cation	H_2O	e =	Cation	Anion	Complex or compound	Compound or element
0.00	At_2		$4OH^-$			-2e		$2AtO^-$		$2H_2O$
+0.20			$2At^-$			-2e				At_2
+0.50		AtO^-	$4OH^-$			-4e		AtO_3^-	$2H_2O$	
+0.70	At_2				$2H_2O$	-2e	$2H^+$		$2HAtO$	
+1.40			$HAtO$		$2H_2O$	-4e	$4H^+$		$HAtO_3$	

EMF	Elem	Anion	Compound or anion	Cation	H_2O	e =	Cation	Anion	Complex or compound	Compound or element
-0.67	Au	$2CN^-$				-e			$Au(CN)_2^-$	
+0.40	Au		$2OH^-$			-e		$Au(OH)_2^-$		
+0.56	Au		$4I^-$			-3e			AuI_4^-	
+058	Au		$2I^-$			-e			AuI_2^-	
+0.66	Au		$4CNS^-$			-3e			$Au(CNS)_4^-$	
+0.87	Au		$4Br^-$			-3e			$AuBr_4^-$	
+0.96	Au		$2Br^-$			-e			$AuBr_2^-$	
+1.00	Au		$4Cl^-$			-3e			$AuCl_4^-$	
+1.15	Au		$2Cl^-$			-e			$AuCl_2^-$	
+1.36	2Au				$3H_2O$	-6e	$6H^+$		Au_2O_3	
+1.40				Au^+		-2e	Au^{+3}			
+1.45	Au				$3H_2O$	-3e	$3H^+$		$Au(OH)_3(c)$	
+1.50	Au					-3e	Au^{+3}			
+1.69	Au					-e	Au^+			

EMF	Elem	Anion	Compound or anion	Cation	H_2O	e =	Cation	Anion	Complex or compound	Compound or element
-4.88			BH			-e	H^+			B
-1.79	B		$4OH^-$			-3e		$H_2BO_3^-$	H_2O	
-0.87	B				$3H_2O$	-3e	$3H^+$		H_3BO_3 (c)	
-0.87	B				$3H_2O$	-3e	$3H^+$		H_3BO_3(aq)	
-0.84	2B				$3H_2O$	-6e	$6H^+$		B_2O_3	
-0.79	4B				$7H_2O$	-12e	$14H^+$	$B_4O_7^{-2}$		
-0.54	H		$H_2BO_3^-$			-e			H_3BO_3	
-0.14			B_2H_6			-6e	$6H^+$			2B

EMF	Elem	Anion	Compound or anion	Cation	H_2O	e =	Cation	Anion	Complex or compound	Compound or element
-3.13	Ba		$Edta^{-4}$			-2e			$Ba(Edta)^{-2}$	
-3.10	Ba		$Amac^{-3}$			-2e			$Ba(Amac)^-$	
-3.09	Ba		Nta^{-3}			-2e			$Ba(Nta)^-$	
-3.05	Ba		$P_4O_{12}^{-4}$			-2e			$BaP_4O_{12}^{-2}$	
-3.03	Ba		$Tmta^{-4}$			-2e			$Ba(Tmta)^{-2}$	
-3.03	Ba		$(FeCy)^{-4}$			-2e			$Ba(FeCy)^{-2}$	
-3.00	Ba		$P_3O_9^{-3}$			-2e			$BaP_3O_9^-$	
-2.99	Ba		$2OH^-$		$8H_2O$	-2e			$Ba(OH)_2 \cdot 8H_2O$	
-2.99	Ba		Cit^{-3}			-2e			$Ba(Cit)^-$	
-2.97	Ba		Ox^{-2}			-2e			$Ba(Ox)$	
-2.97	Ba		$S_2O_3^{-2}$			-2e			BaS_2O_3	
-2.97	Ba		Saa^{-3}			-2e			$Ba(Saa)^-$	
-2.95	Ba		$Tart^{-2}$			-2e			$Ba(Tart)$	
-2.94	Ba		Ap^{-2}			-2e			$Ba(Ap)$	
-2.94	Ba		Mal^{-2}			-2e			$Ba(Mal)$	
-2.93	Ba		$Aspa^{-2}$			-2e			$Ba(Aspa)$	
-2.93	Ba		Glu^-			-2e			$Ba(Glu)^+$	
-2.93	Ba		IO_3^-			-2e			$BaIO_3^+$	
-2.93	Ba		NO_3^-			-2e			$BaNO_3^+$	
-2.93	Ba		Pht^{-2}			-2e			$Ba(Pht)$	
-2.93	Ba		Suc^{-2}			-2e			$Ba(Suc)$	
-2.92	Ba		$Alan^-$			-2e			$Ba(Alan)^+$	
-2.92	Ba		ClO_3^-			-2e			$BaClO_3^+$	
-2.92	Ba		Gl^-			-2e			$Ba(Gl)^+$	
-2.92	Ba		$Glac^-$			-2e			$Ba(Glac)^+$	
-2.92	Ba		Gly^-			-2e			$Ba(Gly)^+$	
-2.92	Ba		OH^-			-2e			$BaOH^+$	
-2.92	Ba		Lac^-			-2e			$Ba(Lac)^+$	
-2.91	Ba					-2e	Ba^{+2}			
-2.91	Ba		Ac^-			-2e			$Ba(Ac)^+$	
-2.91	Ba		Bu^-			-2e			$Ba(Bu)^+$	

EMF	Elem	Anion	Compound or anion	Cation	H_2O	e =	Cation	Anion	Complex or compound	Compound or element
-2.91	Ba		Pr^-			-2e			$Ba(Pr)^+$	
-2.81	Ba		$2OH^-$			-2e			$Ba(OH)_2$	
-1.11			BaH_2			-4e	Ba^{+2}			$2H^+$
-0.69			BaH_2			-2e	$2H^+$			Ba

EMF	Elem	Anion	Compound or anion	Cation	H_2O	e =	Cation	Anion	Complex or compound	Compound or element
-2.63	2Be		$6OH^-$			-4e		$Be_2O_3^{-2}$		$3H_2O$
-2.61	Be		$2OH^-$			-2e			H_2O	BeO
-2.03	Be		$Met(a)^-$			-2e			$Be(Met(a))^+$	
-2.02	Be		$Ist(a)^-$			-2e			$Be(Ist(a))^+$	
-1.98	Be		$Met(b)^-$			-2e			$Be(Met(b))^+$	
-1.97	Be		$2Ist(a)^-$			-2e			$Be(Ist(a))_2$	
-1.97	Be		$Ist(b)^-$			-2e			$Be(Ist(b))^+$	
-1.97	Be		$2Met(a)^-$			-2e			$Be(Met(a))_2$	
-1.95	Be		$Trop^-$			-2e			$Be(Trop)^+$	
-1.93	Be		$AcAc^-$			-2e			$Be(AcAc)^+$	
-1.93	Be		$2Met(b)^-$			-2e			$Be(Met(b))_2$	
-1.92	Be		$2Ist(b)^-$			-2e			$Be(Ist(b))_2$	
-1.92	Be		OH^-			-2e			$BeOH^+$	
-1.91	Be		$2Trop^-$			-2e			$Be(Trop)_2$	
-1.90	Be		$2AcAc^-$			-2e			$Be(AcAc)_2$	
-1.85	Be					-2e	Be^{+2}			
-1.83	Be		Cit^{-3}			-2e			$Be(Cit)^-$	
-1.83	Be		F^-			-2e			BeF^+	
-1.79	Be				H_2O	-2e	$2H^+$		BeO	
-1.77	Be		$HCit^{-2}$			-2e			BeHCit	
-1.76	Be		$2F^-$			-2e			BeF_2	
-1.74	Be		H_2Cit^-			-2e			BeH_2Cit^+	
-1.71	Be		Nac^-			-2e			$Be(Nac)^+$	
-0.91	Be				$2H_2O$	-2e	$4H^+$	BeO_2^{-2}		

EMF	Elem	Anion	Compound or anion	Cation	H_2O	e =	Cation	Anion	Complex or compound	Compound or element
-0.80			BiH_3			-3e	$3H^+$			Bi
-0.46	Bi		$6OH^-$			-6e			BiO_3	$3H_2O$
+0.12	Bi		Br^-			-3e			$BiBr^{+2}$	
+0.15	Bi		Cl^-			-3e			$BiCl^{+2}$	
+0.16	Bi		Cl^-		H_2O	-3e	$2H^+$		$BiOCl$	
+0.16	Bi		NO_3^-			-3e			$BiNO_3^{+2}$	
+0.18	Bi		$2Br^-$			-3e			$BiBr_2^+$	
+0.18	Bi		CNS^-			-3e			$Bi(CNS)^{+2}$	
+0.18	Bi		$2CNS^-$			-3e			$Bi(CNS)_2^+$	
+0.19	Bi		$3Br^-$			-3e			$BiBr_3$	
+0.19	Bi		$3Cl^-$			-3e			$BiCl_3$	
+0.19	Bi		$2Cl^-$			-3e			$BiCl_2^+$	
+0.20	Bi		$4Cl^-$			-3e			$BiCl_4^+$	
+0.20	Bi					-3e	Bi^{+3}			
+0.32	Bi				H_2O	-3e	BiO^+			$2H^+$
+0.37	2Bi				$3H_2O$	-6e	$6H^+$		Bi_2O_3	
+1.59				$2BiO^+$	$2H_2O$	-2e	$4H^+$		Bi_2O_4	

EMF	Elem	Anion	Compound or anion	Cation	H$_2$O	e =	Cation	Anion	Complex or compound	Compound or element
-2.40	Bk					-3e	Bk^{+3}			
+1.60				Bk^{+3}		-e	Bk^{+4}			

EMF	Elem	Anion	Compound or anion	Cation	H_2O	e =	Cation	Anion	Complex or compound	Compound or element
-0.51	H		BrO^-			-e			HBrO	
+0.53	H		Br^-			-e			HBr	
+0.61		$6OH^-$	Br^-			-6e		BrO_3^-	$3H_2O$	
+0.76		$2OH^-$	Br^-			-2e		BrO^-	H_2O	
+1.07			$2Br^-$			-2e			Br_2 (l)	
+1.09			$2Br^-$			-2e			Br_2 (aq)	
+1.52	$\frac{1}{2}Br_2$				$3H_2O$	-5e	$6H^+$	BrO_3^-		
+1.60	$\frac{1}{2}Br_2$				H_2O	-e	H^+		HBrO	

EMF	Elem	Anion	Compound or anion	Cation	H_2O	e =	Cation	Anion	Complex or compound	Compound or element
-0.38	H		HCO_3^-			-e			H_2CO_3	
-0.31	2H		CO_3^{-2}			-2e			H_2CO_3	
-0.20			HCOOH (aq)			-2e	$2H^+$		CO_2 (g)	
+0.06			HCHO (aq)		H_2O	-2e	$2H^+$		HCOOH (aq)	
+0.13			CH_4 (g)			-4e	$4H^+$			C
+0.21	C				$2H_2O$	-4e	$4H^+$		CO_2	
+0.48	C				$3H_2O$	-4e	$6H^+$		CO_3^{-2}	
+0.52			C_2H_6 (g)			-2e	$2H^+$		C_2H_4 (g)	
+0.52	C				H_2O	-2e	$2H^+$		CO	
+0.59			CH_4 (g)		H_2O	-2e	$2H^+$		CH_3OH (aq)	
+0.63	C				$2H_2O$	-2e	$2H^+$		HCOOH	
+0.70			$C_6H_4(OH)_2$			-2e	$2H^+$		$C_6H_4O_2$	
+0.73			C_2H_4 (g)			-2e	$2H^+$		C_2H_4 (g)	
+1.18	C		$4Cl^-$	$4H^+$		-4e	$4H^+$		CCl_4	

EMF	Elem	Anion	Compound or anion	Cation	H_2O	e =	Cation	Anion	Complex or compound	Compound or element
-3.80	Ca					-e	Ca^+			
-3.23	Ca		$Data^{-4}$			-2e			$Ca(Data)^{-2}$	
-3.18	Ca		$Edta^{-4}$			-2e			$Ca(Edta)^{-2}$	
-3.13	Ca		$Amac^{-3}$			-2e			$Ca(Amac)^-$	
-3.11	Ca		Hed^{-3}			-2e			$Ca(Hed)^-$	
-3.11	Ca		Nta^{-3}			-2e			$Ca(Nta)^-$	
-3.08	Ca		$Tmta^{-4}$			-2e			$Ca(Tmta)^{-2}$	
-3.04	Ca		Esb^{-3}			-2e			$Ca(Esb)^-$	
-3.03	Ca		Ers^{-3}			-2e			$Ca(Ers)^-$	
-3.03	Ca		Esa^{-3}			-2e			$Ca(Esa)^-$	
-3.03	Ca		Est^{-3}			-2e			$Ca(Est)^-$	
-3.03	Ca		$Ist(b)^-$			-2e			$Ca(Ist(b))^+$	
-3.03	Ca		$Met(b)^-$			-2e			$Ca(Met(b))^+$	
-3.03	Ca		$P_4O_{12}^{-4}$			-2e			$CaP_4O_{12}^{-2}$	
-3.02	Ca		$2OH^-$			-2e			$Ca(OH)_2$	
-3.02	Ca		$2Amac^{-3}$			-2e			$Ca(Amac)_2^{-4}$	
-3.02	Ca		$Ist(a)^-$			-2e			$Ca(Ist(a))^+$	
-3.02	Ca		$Met(a)^-$			-2e			$Ca(Met(a))^+$	
-3.02	Ca		$P_2O_7^{-4}$			-2e			$CaP_2O_7^{-2}$	
-3.01	Ca		$Himda^{-2}$			-2e			$Ca(Himda)$	
-3.01	Ca		Saa^{-3}			-2e			$Ca(Saa)^-$	
-3.01	Ca		$Trop^-$			-2e			$Ca(Trop)^+$	
-3.00	Ca		Coy^-			-2e			$Ca(Coy)^+$	
-2.98	Ca		$2Ist(a)^-$			-2e			$Ca(Ist(a))_2$	
-2.98	Ca		$2Ist(b)^-$			-2e			$Ca(Ist(b))_2$	
-2.98	Ca		$2Met(b)^-$			-2e			$Ca(Met(b))_2$	
-2.97	Ca		$2Met(a)^-$			-2e			$Ca(Met(a))_2$	
-2.97	Ca		$2Nta^{-3}$			-2e			$Ca(Nta)_2^{-4}$	
-2.97	Ca		$P_3O_9^{-3}$			-2e			$CaP_3O_9^-$	
-2.96	Ca		Cit^{-3}			-2e			$Ca(Cit)^-$	
-2.96	Ca		$HCit^{-2}$			-2e			$CaHCit$	
-2.96	Ca		Ox^{-2}			-2e			$Ca(Ox)$	

EMF	Elem	Anion	Compound or anion	Cation	H_2O	e =	Cation	Anion	Complex or compound	Compound or element
-2.96	Ca		2Trop$^-$			-2e			Ca(Trop)$_2$	
-2.95	Ca		2Coy$^-$			-2e			Ca(Coy)$_2$	
-2.94	Ca		HPO$_4$$^{-2}$			-2e			CaHPO$_4$	
-2.94	Ca		SO$_4$$^{-2}$			-2e			CaSO$_4$	
-2.93	Ca		H$_2$Cit$^-$			-2e			CaH$_2$Cit$^+$	
-2.93	Ca		S$_2$O$_3$$^{-2}$			-2e			CaS$_2$O$_3$	
-2.92	Ca		Ap^{-2}			-2e			Ca(Ap)	
-2.92	Ca		Aspa^{-2}			-2e			Ca(Aspa)	
-2.92	Ca		Tart^{-2}			-2e			Ca(Tart)	
-2.91	Ca		Alan$^-$			-2e			Ca(Alan)$^+$	
-2.91	Ca		Gl$^-$			-2e			Ca(Gl)$^+$	
-2.91	Ca		Glgl$^-$			-2e			Ca(Glgl)$^+$	
-2.91	Ca		Glu$^-$			-2e			Ca(Glu)$^+$	
-2.91	Ca		Mal^{-2}			-2e			Ca(Mal)	
-2.91	Ca		OH$^-$			-2e			Ca(OH)$^+$	
-2.90	Ca		Glac$^-$			-2e			Ca(Glac)$^+$	
-2.90	Ca		Gly$^-$			-2e			Ca(Gly)$^+$	
-2.90	Ca		IO$_3$			-2e			CaIO$_3$$^+$	
-2.90	Ca		Pht^{-2}			-2e			Ca(Pht)	
-2.90	Ca		Suc^{-2}			-2e			Ca(Suc)	
-2.89	Ca		Ac$^-$			-2e			Ca(Ac)$^+$	
-2.89	Ca		Bu$^-$			-2e			Ca(Bu)$^+$	
-2.89	Ca		Lac$^-$			-2e			Ca(Lac)$^+$	
-2.89	Ca		Pr$^-$			-2e			Ca(Pr)$^+$	
-2.88	Ca		NO$_3$$^-$			-2e			CaNO$_3$$^+$	
-2.88	Ca		Ph			-2e			Ca(Ph)$^{+2}$	
-2.87	Ca					-2e	Ca^{+2}			
-2.87	Ca		Sal^{-2}			-2e			Ca(Sal)	
-2.86	Ca		Nac$^-$			-2e			Ca(Nac)$^+$	
-1.90	Ca				H_2O	-2e	2H$^+$		CaO	
-1.04			CaH$_2$			-4e	Ca^{+2}			2H$^+$
+0.78			CaH$_2$			-2e	2H$^+$			Ca

EMF	Elem	Anion	Compound or anion	Cation	H_2O	e =	Cation	Anion	Complex or compound	Compound or element
-2.42			CdH			-e	H^+			Cd
-1.18	Cd		S^{-2}			-2e			CdS	
-1.03	Cd		$4CN^-$			-2e			$Cd(CN)_4^{-2}$	
-0.97	Cd		$Data^{-4}$			-2e			$Cd(Data)^{-2}$	
-0.89	Cd		$Edta^{-4}$			-2e			$Cd(Edta)^{-2}$	
-0.81	Cd		$2OH^-$			-2e			$Cd(OH)_2$	
-0.78	Cd		Hed^{-3}			-2e			$Cd(Hed)^-$	
-0.76	Cd		Tate			-2e			$Cd(Tate)^{+2}$	
-0.74	Cd		CO_3^{-2}			-2e			$CdCO_3$	
-0.72	Cd		Teta			-2e			$Cd(Teta)^{+2}$	
-0.68	Cd		Nta^{-3}			-2e			$Cd(Nta)^-$	
-0.68	Cd		$Oxin^-$			-2e			$Cd(Oxin)^+$	
-0.65	Cd		Deta			-2e			$Cd(Deta)^{+2}$	
-0.63	Cd		$2Oxin^-$			-2e			$Cd(Oxin)_2$	
-0.63	2Cd		$(FeCy)^{-4}$			-4e			$Cd_2(FeCy)$	
-0.62	Cd		$Ndap^{-3}$			-2e			$Cd(Ndap)^-$	
-0.61	Cd		$4NH_3(aq)$			-2e			$Cd(NH_3)_4^{+2}$	
-0.61	Cd		$Himda^{-2}$			-2e			$Cd(Himda)$	
-0.60	Cd		$2Amac^{-3}$			-2e			$Cd(Amac)_2^{-4}$	
-0.60	Cd		Coy^-			-2e			$Cd(Coy)^+$	
-0.59	Cd		Ptn			-2e			$Cd(Ptn)^{+2}$	
-0.57	Cd		$Ndpa^{-3}$			-2e			$Cd(Ndpa)^-$	
-0.57	Cd		$2Nta^{-3}$			-2e			$Cd(Nta)_2^{-4}$	
-0.56	Cd		CN^-			-2e			$CdCN^+$	
-0.56	Cd		2Deta			-2e			$Cd(Deta)_2^{+2}$	
-0.56	Cd		En			-2e			$Cd(En)^{+2}$	
-0.56	Cd		$Imda^{-2}$			-2e			$Cd(Imda)$	
-0.55	Cd		$2CN^-$			-2e			$Cd(CN)_2$	
-0.55	Cd		$2Himda^{-2}$			-2e			$Cd(Himda)_2^{+2}$	
-0.54	Cd		$3CN^-$			-2e			$Cd(CN)_3^-$	
-0.54	Cd		$2Coy^-$			-2e			$Cd(Coy)_2$	
-0.54	Cd		Dge^-			-2e			$Cd(Dge)^+$	

EMF	Elem	Anion	Compound or anion	Cation	H_2O	e =	Cation	Anion	Complex or compound	Compound or element
-0.53	Cd		$Aspa^{-2}$			-2e				$Cd(Aspa)$
-0.53	Cd		$2Em$			-2e				$Cd(Em)_2^{+2}$
-0.53	Cd		$Impa^{-2}$			-2e				$Cd(Impa)$
-0.53	Cd		Pn			-2e				$Cd(Pn)^{+2}$
-0.53	Cd		$2Pn$			-2e				$Cd(Pn)_2^{+2}$
-0.52	Cd		Cit^{-3}			-2e				$Cd(Cit)^-$
-0.52	Cd		$2Imda^{-2}$			-2e				$Cd(Imda)_2^{-2}$
-0.52	Cd		$S_2O_3^{-2}$			-2e				CdS_2O_3
-0.51	Cd		$AcAc^-$			-2e				$Cd(AcAc)^+$
-0.51	Cd		$4CN^-$			-2e				$Cd(CN)_4^{-2}$
-0.50	Cd		$2Dge^-$			-2e				$Cd(Dge)_2$
-0.50	Cd		$Imdp^{-2}$			-2e				$Cd(Imdp)$
-0.50	Cd		Mal^{-2}			-2e				$Cd(Mal)$
-0.50	Cd		Ntp^{-3}			-2e				$Cd(Ntp)^-$
-0.50	Cd		Ox^{-2}			-2e				$Cd(Ox)$
-0.49	Cd		$2Aspa^{-2}$			-2e				$Cd(Aspa)_2^{-2}$
-0.49	Cd		$2Impa^{-2}$			-2e				$Cd(Impa)_2^{-2}$
-0.48	Cd		$2AcAc^-$			-2e				$Cd(AcAc)_2$
-0.48	Cd		Im			-2e				$Cd(Im)^{+2}$
-0.48	Cd		CH_3NH_2			-2e				$Cd(CH_3NH_2)^{+2}$
-0.48	Cd		NH_3			-2e				$Cd(NH_3)^{+2}$
-0.47	Cd		OH^-			-2e				$CdOH^+$
-0.47	Cd		I^-			-2e				CdI^+
-0.47	Cd		$3I^-$			-2e				CdI_3^-
-0.47	Cd		SO_4^{-2}			-2e				$CdSO_4$
-0.47	Cd		$2S_2O_3^{-2}$			-2e				$Cd(S_2O_3)_2^{-2}$
-0.46	Cd		$2CH_3NH_2$			-2e				$Cd(CH_3NH_2)_2^{+2}$
-0.46	Cd		$3Em$			-2e				$Cd(Em)_3^{+2}$
-0.46	Cd		$2Im$			-2e				$Cd(Im)_2^{+2}$
-0.46	Cd		$2NH_3$			-2e				$Cd(NH_3)_2^{+2}$

EMF	Elem	Anion	Compound or anion	Cation	H_2O	e =	Cation	Anion	Complex or compound	Compound or element
-0.46	Cd		3Pn			-2e			$Cd(Pn)_3^{+2}$	
-0.45	Cd		Ac^-			-2e			$Cd(Ac)^+$	
-0.45	Cd		Br^-			-2e			$CdBr^+$	
-0.45	Cd		CSN_2H_4			-2e			$Cd(CSN_2H_4)^{+2}$	
-0.45	Cd		$2I^-$			-2e			CdI_2	
-0.45	Cd		3Im			-2e			$Cd(Im)_3^{+2}$	
-0.45	Cd		$20x^{-2}$			-2e			$Cd(Ox)_2^{-2}$	
-0.44	Cd		CNS^-			-2e			$CdCNS^+$	
-0.44	Cd		$4I^-$			-2e			CdI_4^{-2}	
-0.44	Cd		$3NH_3$			-2e			$Cd(NH_3)_3^{+2}$	
-0.43	Cd		$2Ac^-$			-2e			$Cd(Ac)_2$	
-0.43	Cd		$3Br^-$			-2e			$CdBr_3^-$	
-0.43	Cd		$3CH_3NH_2$			-2e			$Cd(CH_3NH_2)_3^{+2}$	
-0.43	Cd		$4CNS^-$			-2e			$Cd(CNS)_4^{-2}$	
-0.43	Cd		$2CSN_2H_4$			-2e			$Cd(CSN_2H_4)_2^{+2}$	
-0.43	Cd		4Im			-2e			$Cd(Im)_4^{+2}$	
-0.43	Cd		$4NH_3$			-2e			$Cd(NH_3)_4^{+2}$	
-0.42	Cd		$2Br^-$			-2e			$CdBr_2$	
-0.42	Cd		$4CH_3NH_2$			-2e			$Cd(CH_3NH_2)_4^{+2}$	
-0.42	Cd		$2Cl^-$			-2e			$CdCl_2$	
-0.42	Cd		$2CNS^-$			-2e			$Cd(CNS)_2$	
-0.42	Cd		$3CNS^-$			-2e			$Cd(CNS)_3^-$	
-0.41	Cd		$4Br^-$			-2e			$CdBr_4^{-2}$	
-0.41	Cd		$3Cl^-$			-2e			$CdCl_3^-$	
-0.41	Cd		Nac^-			-2e			$Cd(Nac)^+$	
-0.41	Cd		NO_3^-			-2e			$CdNO_3^+$	
-0.40	Cd					-2e	Cd^{+2}			

EMF	Elem	Anion	Compound or anion	Cation	H_2O	e =	Cation	Anion	Complex or compound	Compound or element
-0.39	Cd		$5NH_3$			-2e			$Cd(NH_3)_5^{+2}$	
-0.37	Cd		$3CNS^-$			-2e			$Cd(CNS)_3^-$	
-0.35	Cd		$6NH_3$			-2e			$Cd(NH_3)_6^{+2}$	
-0.35			$Cd \cdot Hg$			-2e	Cd^{+2}			Hg
+0.06	Cd				H_2O	-2e	$2H^+$		CdO	

EMF	Elem	Anion	Compound or anion	Cation	H_2O	e =	Cation	Anion	Complex or compound	Compound or element
-2.87	Ce		$3OH^-$			-3e			$Ce(OH)_3$	
-2.82	Ce		$P_2O_7^{-4}$			-3e			$CeP_2O_7^-$	
-2.81	Ce		$Data^{-4}$			-3e			$Ce(Data)^-$	
-2.78	Ce		$Edta^{-4}$			-3e			$Ce(Edta)^-$	
-2.77	Ce		OH^-			-3e			$CeOH^{+2}$	
-2.69	2Ce		$3S^{-2}$			-6e			Ce_2S_3	
-2.68	Ce		$2Amac^{-3}$			-3e			$Ce(Amac)_2^{-3}$	
-2.66	Ce		$Oxin^-$			-3e			$Ce(Oxin)^{+2}$	
-2.64	Ce		Nta^{-3}			-3e			$Ce(Nta)$	
-2.64	Ce		$2Oxin^-$			-3e			$Ce(Oxin)_2^+$	
-2.63	Ce		Oxd^-			-3e			$Ce(Oxd)^{+2}$	
-2.61	Ce		Ox^{-2}			-3e			$Ce(Ox)^+$	
-2.58	Ce		$AcAc^-$			-3e			$Ce(AcAc)^{+2}$	
-2.56	Ce		$2AcAc^-$			-3e			$Ce(AcAc)_2^+$	
-2.56	Ce		$2Ox^{-2}$			-3e			$Ce(Ox)_2^-$	
-2.55	Ce		SO_4^{-2}			-3e			$CeSO_4^+$	
-2.54	Ce		F^-			-3e			CeF^{+2}	
-2.52	Ce		SO_4^{-2}			-3e			$CeSO_4^+$	
-2.51	Ce		Ac^-			-3e			$Ce(Ac)^{+2}$	
-2.50	Ce		$2Ac^-$			-3e			$Ce(Ac)_2^+$	
-2.50	Ce		$3Ox^{-2}$			-3e			$Ce(Ox)_3^{-3}$	
-2.49	Ce		$3Ac^-$			-3e			$Ce(Ac)_3$	
-2.49	Ce		Br^-			-3e			$CeBr^{+2}$	
-2.48	Ce					-3e	Ce^{+3}			
-2.45	Ce		$3AcAc^-$			-3e			$Ce(AcAc)_3$	
+1.61				Ce^{+3}		-e	Ce^{+4}			

EMF	Elem	Anion	Compound or anion	Cation	H_2O	e =	Cation	Anion	Complex or compound	Compound or element
-2.10	Cf					-3e	Cf^{+3}			

EMF	Elem	Anion	Compound or anion	Cation	H_2O	e =	Cation	Anion	Complex or compound	Compound or element
-0.43	H		ClO^-			-e			$HClO$	
-0.12	H		ClO_2^-			-e			$HClO_2$	
+0.06	H		ClO_3^-			-e			$HClO_3$	
+0.33		$2OH^-$	ClO_2^-			-2e		ClO_3^-		H_2O
+0.36		$2OH^-$	ClO_3^-			-2e		ClO_4^-		H_2O
+0.41	H		Cl^-			-e			HCl	
+0.47	H		ClO_4^-			-e			$HClO_4$	
+0.66		$2OH^-$	ClO^-			-2e		ClO_2^-	H_2O	
+0.89		$2OH^-$	Cl^-			-2e			ClO^-	H_2O
+0.99			$2HCl$			-2e	$2H^+$				Cl_2
+1.16			ClO_2^-			-e			ClO_2 (g)	
+1.19			ClO_3^-		H_2O	-2e	$2H^+$	ClO_4^-		
+1.21			$HClO_2$		H_2O	-2e	$3H^+$	ClO_3^-		
+1.28			$HClO_2$			-e	H^+		ClO_2	
+1.36			$2Cl^-$			-2e				Cl_2
+1.63	$\frac{1}{2}Cl_2$				H_2O	-e	H^+		$HClO$	
+1.65			$HClO$		H_2O	-2e	$2H^+$		$HClO_2$	
+1.71	Cl_2				H_2O	-2e	$2H^+$		Cl_2O	

EMF	Elem	Anion	Compound or anion	Cation	H_2O	e =	Cation	Anion	Complex or compound	Compound or element
-2.70	Cm					-3e	Cm^{+3}			

EMF	Elem	Anion	Compound or anion	Cation	H_2O	e =	Cation	Anion	Complex or compound	Compound or element
-0.84	Co		$Data^{-4}$			-2e			$Co(Data)^{-2}$	
-0.76	Co		$Edta^{-4}$			-2e			$Co(Edta)^{-2}$	
-0.73	Co		$2OH^-$			-2e			$Co(OH)_2$	
-0.71	Co		Hed^{-3}			-2e			$Co(Hed)^-$	
-0.66	Co		Tate			-2e			$Co(Tate)^{+2}$	
-0.64	Co		CO_3^{-2}			-2e			$CoCO_3$	
-0.61	Co		Nta^{-3}			-2e			$Co(Nta)^-$	
-0.61	Co		Teta			-2e			$Co(Teta)^{+2}$	
-0.59	Co		$Oxin^-$			-2e			$Co(Oxin)^+$	
-0.58	Co		$Ndap^{-3}$			-2e			$Co(Ndap)^-$	
-0.57	Co		Oxd^-			-2e			$Co(Oxd)^+$	
-0.55	Co		$2Oxin^-$			-2e			$Co(Oxin)_2$	
-0.54	Co		$2Oxd^-$			-2e			$Co(Oxd)_2$	
-0.52	Co		Deta			-2e			$Co(Deta)^{+2}$	
-0.52	Co		$Himda^{-2}$			-2e			$Co(Himda)$	
-0.52	Co		$Ist(a)^-$			-2e			$Co(Ist(a))^+$	
-0.52	Co		$Met(a)^-$			-2e			$Co(Met(a))^+$	
-0.51	Co		$Ist(b)^-$			-2e			$Co(Ist(b))^+$	
-0.51	Co		$Met(b)^-$			-2e			$Co(Met(b))^+$	
-0.51	Co		$Ndpa^{-3}$			-2e			$Co(Ndpa)^-$	
-0.49	Co		$Imda^{-2}$			-2e			$Co(Imda)$	
-0.49	Co		$Trop^-$			-2e			$Co(Trop)^+$	
-0.48	Co		Coy^-			-2e			$Co(Coy)^+$	
-0.48	Co		$2Ist(a)^-$			-2e			$Co(Ist(a))_2$	
-0.48	Co		Ptn			-2e			$Co(Ptn)^{+2}$	
-0.47	Co		$2Ist(b)^-$			-2e			$Co(Ist(b))_2$	
-0.47	Co		$2Met(a)^-$			-2e			$Co(Met(a))_2$	
-0.46	Co		2Deta			-2e			$Co(Deta)_2^{+2}$	
-0.46	Co		$Impa^{-2}$			-2e			$Co(Impa)$	
-0.46	Co		$2Met(b)^-$			-2e			$Co(Met(b))_2$	
-0.45	Co		Af^-			-2e			$Co(Af)^+$	
-0.45	Co		$Aspa^{-2}$			-2e			$Co(Aspa)$	

EMF	Elem	Anion	Compound or anion	Cation	H_2O	e =	Cation	Anion	Complex or compound	Compound or element
-0.45	Co		En			-2e			$Co(En)^{+2}$	
-0.45	Co		$2Trop^-$			-2e			$Co(Trop)_2$	
-0.44	Co		$AcAc^-$			-2e			$Co(AcAc)^+$	
-0.44	Co		Dge^-			-2e			$Co(Dge)^+$	
-0.44	Co		$2Imda^{-2}$			-2e			$Co(Imda)_2^{-2}$	
-0.43	Co		$2Coy^-$			-2e			$Co(Coy)_2$	
-0.43	Co		Gl^-			-2e			$Co(Gl)^+$	
-0.43	Co		$Imdp^{-2}$			-2e			$Co(Imdp)$	
-0.42	Co		$2Af^-$			-2e			$Co(Af)_2$	
-0.42	Co		$Alan^-$			-2e			$Co(Alan)^+$	
-0.42	Co		$2En$			-2e			$Co(En)_2^{+2}$	
-0.42	Co		Ntp^{-3}			-2e			$Co(Ntp)^-$	
-0.42	Co		Ox^{-2}			-2e			$Co(Ox)$	
-0.42	Co		$Sald^-$			-2e			$Co(Sald)^+$	
-0.41	Co		$2Aspa^{-2}$			-2e			$Co(Aspa)_2^{-2}$	
-0.41	Co		$2Himda^{-2}$			-2e			$Co(Himda)_2^{-2}$	
-U.41	Cu		$2Impa^{-2}$			-2e			$Co(Impa)_2^{-2}$	
-0.41	Co		OH^-			-2e			$CoOH^+$	
-0.40	Co		$2AcAc^-$			-2e			$Co(AcAc)_2$	
-0.40	Co		$2Gl^-$			-2e			$Co(Gl)_2$	
-0.40	Co		$2Nta^{-3}$			-2e			$Co(Nta)_2^{-4}$	
-0.39	Co		$2Alan^-$			-2e			$Co(Alan)_2$	
-0.39	Co		$3Ist(b)^-$			-2e			$Co(Ist(b))_3^-$	
-0.39	Co		Mal^{-2}			-2e			$Co(Mal)$	
-0.39	Co		$2Sald^-$			-2e			$Co(Sald)_2$	
-0.39	Co		$3Trop^-$			-2e			$Co(Trop)_3^-$	
-0.38	Co		$2Dge^-$			-2e			$Co(Dge)_2$	
-0.38	Co		$Glgl^-$			-2e			$Co(Glgl)^+$	
-0.38	Co		$2Imdp^{-2}$			-2e			$Co(Imdp)_2^{-2}$	
-0.38	Co		$SSald^{-2}$			-2e			$Co(SSald)$	
-0.37	Co		$2Amac^{-3}$			-2e			$Co(Amac)_2^{-4}$	
-0.37	Co		$3En$			-2e			$Co(En)_3^{+2}$	

EMF	Elem	Anion	Compound or anion	Cation	H_2O	e =	Cation	Anion	Complex or compound	Compound or element
-0.35	Co		$4CNS^-$			-2e			$Co(CNS)_4^{+2}$	
-0.35	Co		$2Glgl^-$			-2e			$Co(Glgl)_2$	
-0.35	Co		$2Ox^{-2}$			-2e			$Co(Ox)_2^{-2}$	
-0.35	Co		SO_4^{-2}			-2e			$CoSO_4$	
-0.34	Co		NH_3			-2e			$Co(NH_3)^{+2}$	
-0.34	Co		$S_2O_3^{-2}$			-2e			CoS_2O_3	
-0.33	Co		$2NH_3$			-2e			$Co(NH_3)_2^{+2}$	
-0.33	Co		Pht^{-2}			-2e			$Co(Pht)$	
-0.31	Co		$3NH_3$			-2e			$Co(NH_3)_3^{+2}$	
-0.30	Co		$3CNS^-$			-2e			$Co(CNS)_3^-$	
-0.30	Co		$4NH_3$			-2e			$Co(NH_3)_4^{+2}$	
-0.29	Co		$5NH_3$			-2e			$Co(NH_3)_5^{+2}$	
-0.28	Co					-2e	Co^{+2}			
-0.28	Co		CNS^-			-2e			$CoCNS^+$	
-0.28	Co		Nac^-			-2e			$Co(Nac)^+$	
-0.28	Co		Suc^{-2}			-2e			$Co(Suc)$	
-0.26	Co		$2CNS^-$			-2e			$Co(CNS)_2$	
-0.26	Co		$6NH_3$			-2e			$Co(NH_3)_6^{+2}$	
+0.11			$Co(NH_3)_6^{+2}$			-e			$Co(NH_3)_6^{+3}$	
+0.17		OH^-	$Co(OH)_2$			-e			$Co(OH)_3$	
+0.27	2Co		$(FeCy)^{-4}$			-4e			$Co_2(FeCy)$	
+1.81				Co^{+2}		-e	Co^{+3}			

EMF	Elem	Anion	Compound or anion	Cation	H_2O	e =	Cation	Anion	Complex or compound	Compound or element
-1.48	Cr		$3OH^-$			-3e			$Cr(OH)_3(c)$	
-1.34	Cr		$3OH^-$			-3e			$Cr(OH)_3(hyd)$	
-1.27	Cr		$4OH^-$			-3e			CrO_2	$2H_2O$
-0.94	Cr		OH^-			-3e			$CrOH^{+2}$	
-0.91	Cr					-2e	Cr^{+2}			
-0.83	Cr		F^-			-3e			CrF^{+2}	
-0.81	Cr		$2F^-$			-3e			CrF_2^+	
-0.79	Cr		$3F^-$			-3e			CrF_3	
-0.78	Cr		CNS^-			-3e			$CrCNS^{+2}$	
-0.76	Cr		$2CNS^-$			-3e			$Cr(CNS)_2^+$	
-0.58	2Cr				$3H_2O$	-6e	$6H^+$		Cr_2O_3	
-0.74	Cr					-3e	Cr^{+3}			
-0.41				Cr^{+2}		-e	Cr^{+3}			
-0.19	2H		CrO_4^{-2}			-2e			H_2CrO_4	
-0.13		$5OH^-$	$Cr(OH)_3$			-3e		CrO_4^{-2}	$4H_2O$	
+0.06	H		$HCrO_4^-$			-e			H_2CrO_4	
+1.33				$2Cr^{+3}$	$7H_2O$	-6e	$14H^+$	$Cr_2O_7^{-2}$		

EMF	Elem	Anion	Compound or anion	Cation	H_2O	e =	Cation	Anion	Complex or compound	Compound or element
-2.92	Cs					-e	Cs^+			
-1.30			CsH			-2e	Cs^+			H^+
+0.32			CsH			-e	H^+			Cs

EMF	Elem	Anion	Compound or anion	Cation	H_2O	e =	Cation	Anion	Complex or compound	Compound or element
-2.78			CuH			-e	H^+			Cu
-1.28	Cu		$4CN^-$			-e			$Cu(CN)_4^{-3}$	
-1.17	Cu		$3CN^-$			-e			$Cu(CN)_3^{-2}$	
-0.89	2Cu		S^{-2}			-2e			Cu_2S	
-0.70	Cu		S^{-2}			-2e			CuS	
-0.64	Cu		CN^-			-e			CuCN	
-0.43	Cu		$2CN^-$			-e			$Cu(CN)_2^-$	
-0.36	2Cu		$2OH^-$			-2e			Cu_2O	H_2O
-0.29	Cu		$Data^{-4}$			-2e			$Cu(Data)^{-2}$	
-0.27	Cu		CNS^-			-e			CuCNS	
-0.27	Cu		2(12DAP)			-2e			$Cu(12DAP)_2^{+2}$	
-0.26	Cu		$2OH^-$			-2e			CuO	H_2O
-0.26	Cu		Teta			-2e			$Cu(Teta)^{+2}$	
-0.23	Cu		$2Glu^{-2}$			-2e			$Cu(Glu)_2^{-2}$	
-0.22	Cu		NH_3			-2e			$Cu(NH_3)^{+2}$	
-0.22	Cu		$2OH^-$			-2e			$Cu(OH)_2$	
-0.22	Cu		$Edta^{-4}$			-2e			$Cu(Edta)^{-2}$	
-0.22	Cu		Tate			-2e			$Cu(Tate)^{+2}$	
-0.19	Cu		I^-			-e			CuI	
-0.17	Cu		Hed^{-3}			-2e			$Cu(Hed)^-$	
-0.16	Cu		2(13DAP)			-2e			$Cu(13DAP)_2^{+2}$	
-0.13	2Cu		$(FeCy)^{-4}$			-4e			$Cu_2(FeCy)$	
-0.13	Cu		Deta			-2e			$Cu(Deta)^{+2}$	
-0.12	Cu		$2NH_3$			-e			$Cu(NH_3)_2^+$	
-0.12	Cu		2En			-e			$Cu(En)_2^+$	
-0.09	Cu		$S_2O_3^{-2}$			-e			$CuS_2O_3^-$	
-0.08		$2OH^-$	Cu_2O		H_2O	-2e			$2Cu(OH)_2$	
-0.08	Cu		$2IO_3^-$			-2e			$Cu(IO_3)_2$	
-0.08	Cu		Cit^{-3}			-2e			$Cu(Cit)^-$	
-0.06	Cu		$mOxin^-$			-2e			$Cu(mOxin)^+$	
-0.06	Cu		$Oxin^-$			-2e			$Cu(Oxin)^+$	
-0.04	Cu		$2Oxin^-$			-2e			$Cu(Oxin)_2$	

EMF	Elem	Anion	Compound or anion	Cation	H_2O	e =	Cation	Anion	Complex or compound	Compound or element
-0.03	Cu		2mOxin$^-$			-2e			Cu(mOxin)$_2$	
-0.02	Cu		4SCN$^-$			-e			Cu(SCN)$_4^{-3}$	
-0.02	Cu		2S$_2$O$_3^{-2}$			-2e			Cu(S$_2$O$_3$)$_2^{-2}$	
-0.02	Cu		Glu^{-2}			-2e			Cu(Glu)	
-0.01	Cu		Ndap^{-3}			-2e			Cu(Ndap)$^-$	
+0.01	Cu		Ptn			-2e			Cu(Ptn)$^{+2}$	
+0.02	Cu		12DAP			-2e			Cu(12DAP)$^{+2}$	
+0.03	Cu		Br$^-$			-e			CuBr	
+0.03	Cu		En			-2e			Cu(En)$^{+2}$	
+0.03	Cu		Has$^-$			-2e			Cu(Has)$^+$	
+0.03	Cu		N$_3^-$			-e			CuN$_3$	
+0.03	Cu		3NH$_3$			-2e			Cu(NH$_3$)$_3^{+2}$	
+0.03	Cu		Imda^{-2}			-2e			Cu(Imda)	
+0.03	Cu		Impa^{-2}			-2e			Cu(Impa)	
+0.03	Cu		Pn			-2e			Cu(Pn)$^{+2}$	
+0.03	Cu		Sal^{-2}			-2e			Cu(Sal)	
+0.04	Cu		DmHas$^-$			-2e			Cu(DmHas)$^+$	
+0.04	Cu		Nta^{-3}			-2e			Cu(Nta)$^-$	
+0.04	Cu		Oxd$^-$			-2e			Cu(Oxd)$^+$	
+0.04	Cu		Tmen			-2e			Cu(Tmen)$^{+2}$	
+0.05	Cu		2DmHas$^-$			-2e			Cu(DmHas)$_2$	
+0.05	Cu		Hox$^-$			-2e			Cu(Hox)$^+$	
+0.05	Cu		13DAP			-2e			Cu(13DAP)$^{+2}$	
+0.06	Cu		Cin$^-$			-2e			Cu(Cin)$^+$	
+0.06	Cu		Coy$^-$			-2e			Cu(Coy)$^+$	
+0.06	Cu		2Has$^-$			-2e			Cu(Has)$_2$	
+0.06	Cu		Imdp^{-2}			-2e			Cu(Imdp)	
+0.06	Cu		2Oxd$^-$			-2e			Cu(Oxd)$_2$	
+0.07	Cu		Af$^-$			-2e			Cu(Af)$^+$	
+0.07	Cu		2En			-2e			Cu(En)$_2^{+2}$	
+0.07	Cu		Ntp^{-3}			-2e			Cu(Ntp)$^-$	
+0.07	Cu		2Pn			-2e			Cu(Pn)$_2^{+2}$	

EMF	Elem	Anion	Compound or anion	Cation	H_2O	e =	Cation	Anion	Complex or compound	Compound or element
+0.07	Cu		2Ptn			-2e			$Cu(Ptn)_2^{+2}$	
+0.08	Cu		$2Hox^-$			-2e			$Cu(Hox)_2$	
+0.08	Cu		$MpHas^-$			-2e			$Cu(MpHas)^+$	
+0.08	Cu		SO_3^{-2}			-e			$CuSO_3^-$	
+0.09	Cu		$AcAc^-$			-2e			$Cu(AcAc)^+$	
+0.09	Cu		$2Af^-$			-2e			$Cu(Af)_2$	
+0.09	Cu		$Alan^-$			-2e			$Cu(Alan)^+$	
+0.09	Cu		$Aspa^{-2}$			-2e			$Cu(Aspa)$	
+0.09	Cu		$2Cin^-$			-2e			$Cu(Cin)_2$	
+0.09	Cu		Gl^-			-2e			$Cu(Gl)^+$	
+0.09	Cu		$2MpHas^-$			-2e			$Cu(MpHas)_2$	
+0.10	Cu		Dge^-			-2e			$Cu(Dge)^+$	
+0.11	Cu		$2Trop^-$			-2e			$Cu(Trop)_2$	
+0.12	Cu		$Sald^-$			-2e			$Cu(Sald)^+$	
+0.13	Cu		$2Coy^-$			-2e			$Cu(Coy)_2$	
+0.13	Cu		2Tmen			-2e			$Cu(Tmen)_2^{+2}$	
+0.14	Cu		$2AcAc^-$			-2e			$Cu(AcAc)_2$	
+0.14	Cu		$2Alan^-$			-2e			$Cu(Alan)_2$	
+0.14	Cu		$2Aspa^{-2}$			-2e			$Cu(Aspa)_2^{-2}$	
+0.14	Cu		Cl^-			-e			CuCl	
+0.14	Cu		$2Gl^-$			-2e			$Cu(Gl)_2$	
+0.14	Cu		$P_2O_7^{-4}$			-2e			$CuP_2O_7^{-2}$	
+0.15	Cu		$2(DMMal)^{-4}$			-2e			$Cu(DMMal)_2^{-2}$	
+0.15				Cu^+		-e	Cu^{+2}			
+0.15	Cu		NH_3			-e			$CuNH_3^+$	
+0.15	Cu		OH^-			-2e			$CuOH^+$	
+0.15	Cu		$2Sal^{-2}$			-2e			$Cu(Sal)_2^{-2}$	
+0.15	Cu		Tf^-			-2e			$Cu(Tf)^+$	
+0.16	Cu		$Glgl^-$			-2e			$Cu(Glgl)^+$	
+0.16	Cu		Ox^{-2}			-2e			$Cu(Ox)$	
+0.17	Cu		$2Glgl^-$			-2e			$Cu(Glgl)_2$	
+0.17	Cu		$2Br^-$			-e			$CuBr_2^-$	

EMF	Elem	Anion	Compound or anion	Cation	H_2O	e =	Cation	Anion	Complex or compound	Compound or element
+0.17	Cu		$2Imda^{-2}$			-2e			$Cu(Imda)_2^{-2}$	
+0.17	Cu		HPO_4^{-2}			-2e			$CuHPO_4$	
+0.17	Cu		Mal^{-2}			-2e			$Cu(Mal)$	
+0.17	Cu		$2Sald^-$			-2e			$Cu(Sald)_2$	
+0.18	Cu		$2Deta$			-2e			$Cu(Deta)_2^{+2}$	
+0.18	Cu		$SSald^{-2}$			-2e			$Cu(SSald)$	
+0.19	Cu		$2Dge^-$			-2e			$Cu(Dge)_2$	
+0.19	Cu		$EtMal^{-2}$			-2e			$Cu(EtMal)$	
+0.20			Cu_2O	$2H^+$		-2e	$2Cu^{+2}$			H_2O
+0.20	Cu		$2H_2PO_4^-$			-2e			$Cu(H_2PO_4)_2$	
+0.20	Cu		$DMMal^{-2}$			-2e			$Cu(DMMal)$	
+0.21	Cu		$2Himda^{-2}$			-2e			$Cu(Himda)_2^{-2}$	
+0.21	Cu		Im			-2e			$Cu(Im)^{+2}$	
+0.21	Cu		$2Impa^{-2}$			-2e			$Cu(Impa)_2^{-2}$	
+0.22	Cu		$2SSald^{-2}$			-2e			$Cu(SSald)_2^{-2}$	
+0.22	Cu		Bu^-			-2e			$Cu(Bu)^+$	
+0.23	Cu		$2Im$			-2e			$Cu(Im)_2^{+2}$	
+0.23	Cu		$2Imdp^{-2}$			-2e			$Cu(Imdp)_2^{-2}$	
+0.24	Cu		$2NH_3$			-2e			$Cu(NH_3)_2^{+2}$	
+0.24	Cu		$Adip^{-2}$			-2e			$Cu(Adip)$	
+0.24	Cu		Suc^{-2}			-2e			$Cu(Suc)$	
+0.25	Cu		$3CN^-$			-e			$Cu(CN)_3^{-2}$	
+0.25	Cu		$P_4O_{12}^{-4}$			-2e			$CuP_4O_{12}^{-2}$	
+0.25	Cu		$Tart^{-2}$			-2e			$Cu(Tart)$	
+0.26	Cu		$3Im$			-2e			$Cu(Im)_3^{+2}$	
+0.26	Cu		Gly^{-2}			-2e			$Cu(Gly)$	
+0.27	Cu		Ac^-			-2e			$Cu(Ac)^+$	
+0.27	Cu		$2P_2O_7^{-4}$			-2e			$Cu(P_2O_7)_2^{-6}$	
+0.27	Cu		Pyr			-2e			$Cu(Pyr)^{+2}$	

EMF	Elem	Anion	Compound or anion	Cation	H_2O	e =	Cation	Anion	Complex or compound	Compound or element
+0.27	Cu		SO_4^{-2}			-2e				$CuSO_4$
+0.27	Cu		$FumA^{-2}$			-2e				$Cu(FumA)$
+0.28	Cu		4Im			-2e				$Cu(Im)_4^{+2}$
+0.28	Cu		$4NH_3$			-2e				$Cu(NH_3)_4^{+2}$
+0.28	Cu		2Pyr			-2e				$Cu(Pyr)_2^{+2}$
+0.28	Cu		Ac^-			-e				$Cu(Ac)$
+0.28	Cu		$2Br^-$			-2e				$CuBr_2$
+0.28	Cu		iBu^-			-2e				$Cu(iBu)^+$
+0.28	Cu		CHO^-			-2e				$Cu(CHO)^+$
+0.28	Cu		$HexA^-$			-2e				$Cu(HexA)^+$
+0.28	Cu		$\emptyset Ac^-$			-2e				$Cu(\emptyset Ac)^+$
+0.29	Cu		$BrAc^-$			-2e				$Cu(BrAc)^+$
+0.29	Cu		$ClAc^-$			-2e				$Cu(ClAc)^+$
+0.30	Cu		$2P_4O_{12}^{-4}$			-2e				$Cu(P_4O_{12})_2^{-6}$
+0.30	Cu		3Pyr			-2e				$Cu(Pyr)_3^{+2}$
+0.30	Cu		$BrBu^-$			-2e				$Cu(BrBu)^+$
+0.31	Cu		$2Ac^-$			-2e				$Cu(Ac)_2$
+0.31	Cu		4Pyr			-2e				$Cu(Pyr)_4^{+2}$
+0.31	Cu		NO_2^-			-2e				$CuNO_2^+$
+0.33	Cu		Br^-			-2e				$CuBr^+$
+0.33	Cu		Nac^-			-2e				$Cu(Nac)^+$
+0.33	Cu		SO_4^{-2}			-2e				$CuSO_4$
+0.33	3Cu		$2PO_4^{-3}$			-6e				$Cu_3(PO_4)_2$
+0.34	Cu					-2e	Cu^{+2}			
+0.34	Cu		Cl^-			-2e				$CuCl^+$
+0.35	Cu		$5NH_3$			-2e				$Cu(NH_3)_5^{+2}$
+0.36	Cu		$2Cl^-$			-2e				$CuCl_2$
+0.40	Cu		$2S_2O_3^{-2}$			-e				$Cu(S_2O_3)_2^{-3}$
+0.42	Cu		$4CN^-$			-e				$Cu(CN)_4^{-3}$
+0.42	Cu		$3S_2O_3^{-2}$			-e				$Cu(S_2O_3)_3^{-5}$
+0.45	Cu		3Pyr			-e				$Cu(Pyr)_3^+$

EMF	Elem	Anion	Compound or anion	Cation	H_2O	e =	Cation	Anion	Complex or compound	Compound or element
+0.46	Cu		4Pyr			-e			$Cu(Pyr)_4^+$	
+0.46	Cu		$2SO_3^{-2}$			-e			$Cu(SO_3)_2^{-3}$	
+0.48	Cu		$3SO_3^{-2}$			-e			$Cu(SO_3)_3^{-5}$	
+0.52	Cu					-e	Cu^+			
+0.54			CuCl			-e	Cu^{+2}	Cl^-		
+0.62				Cu^+	H_2O	-e	$2H^+$		CuO	
+0.64			CuBr			-e	Cu^{+2}	Br^-		
+0.67			Cu_2O		H_2O	-2e	$2H^+$			2CuO (a)
+0.75			Cu_2O		H_2O	-2e	$2H^+$			2CuO (b)
+0.86			CuI			-e	Cu^{+2}	I^-		
+1.10			$Cu(CN)_2^-$			-e	Cu^{+2}	$2CN^-$		
+1.13	Cu				$2H_2O$	-2e	$3H^+$	$HCuO_2^-$		
+1.52	Cu				$2H_2O$	-2e	$4H^+$	CuO_2^{-2}		
+1.73				Cu^+	$2H_2O$	-e	$3H^+$	$HCuO_2^-$		
+1.78			Cu_2O		$3H_2O$	-2e	$4H^+$	$2HCuO_2^-$		
+2.51				Cu^+	$2H_2O$	-e	$4H^+$	CuO_2^{-2}		
+2.56			Cu_2O		$3H_2O$	-2e	$6H^+$	$2CuO_2^{-2}$		

EMF	Elem	Anion	Compound or anion	Cation	H_2O	e =	Cation	Anion	Complex or compound	Compound or element
-2.78	Dy		$3OH^-$			-3e			$Dy(OH)_3$	
-2.74	Dy		$Data^{-4}$			-3e			$Dy(Data)^-$	
-2.70	Dy		$Edta^{-4}$			-3e			$Dy(Edta)^-$	
-2.46	Dy		$Oxac^{-2}$			-3e			$Dy(Oxac)^+$	
-2.44	Dy		$2Oxac^{-2}$			-3e			$Dy(Oxac)_2^-$	
-2.35	Dy					-3e	Dy^{+3}			
-1.96	2Dy				$3H_2O$	-6e	$6H^+$		Dy_2O_3	

EMF	Elem	Anion	Compound or anion	Cation	H_2O	e =	Cation	Anion	Complex or compound	Compound or element
-2.75	Er		$3OH^-$			-3e			$Er(OH)_3$	
-2.71	Er		$Data^{-4}$			-3e			$Er(Data)^-$	
-2.65	Er		$Edta^{-4}$			-3e			$Er(Edta)^-$	
-2.37	Er		SO_4^{-2}			-3e			$ErSO_4^+$	
-2.30	Er					-3e	Er^{+3}			
-1.92	2Er				$3H_2O$	-6e	$6H^+$		Er_2O_3	

EMF	Elem	Anion	Compound or anion	Cation	H_2O	e =	Cation	Anion	Complex or compound	Compound or element
-3.40	Eu					-2e	Eu^{+2}			
-2.83	Eu		$3OH^-$			-3e			$Eu(OH)_3$	
-2.78	Eu		$Data^{-4}$			-3e			$Eu(Data)^-$	
-2.74	Eu		$Edta^{-4}$			-3e			$Eu(Edta)^-$	
-2.53	Eu		$AcAc^-$			-3e			$Eu(AcAc)^{+2}$	
-2.50	Eu		$2AcAc^-$			-3e			$Eu(AcAc)_2^+$	
-2.48	Eu		$3AcAc^-$			-3e			$Eu(AcAc)_3$	
-2.41	Eu					-3e	Eu^{+3}			
-2.00	2Eu				$3H_2O$	-6e	$6H^+$		Eu_2O_3	
-0.43				Eu^{+2}		-e	Eu^{+3}			

EMF	Elem	Anion	Compound or anion	Cation	H_2O	e =	Cation	Anion	Complex or compound	Compound or element
-0.19	H		F^-			-e			HF	
+1.44	F_2				H_2O	-2e	$2H^+$		F_2O	
+2.15			$2F^-$		H_2O	-4e	$2H^+$		F_2O	
+2.87			$2F^-$			-2e				F_2 (g)
+3.06			2HF (aq)			-2e	$2H^+$			F_2 (g)

EMF	Elem	Anion	Compound or anion	Cation	H_2O	e =	Cation	Anion	Complex or compound	Compound or element
-1.16	Fe		$6CN^-$			-2e				$(FeCy)^{-4}$
-0.95	Fe		S^{-2}			-2e				FeS (a)
-0.88	Fe		$2OH^-$			-2e				$Fe(OH)_2$
-0.87	Fe		$Edta^{-4}$			-2e				$Fe(Edta)^{-2}$
-0.78	Fe		Hed^{-3}			-2e				$Fe(Hed)^-$
-0.76	Fe		CO_3^{-2}			-2e				$FeCO_3$
-0.72		S^{-2}	2FeS (a)			-2e				Fe_2S_3
-0.70	Fe		Nta^{-3}			-2e				$Fe(Nta)^-$
-0.70	Fe		Tate			-2e				$Fe(Tate)^{+2}$
-0.67	Fe		Teta			-2e				$Fe(Teta)^{+2}$
-0.62	Fe		Deta			-2e				$Fe(Deta)^{+2}$
-0.61	Fe		Ph			-2e				$Fe(Ph)^{+2}$
-0.59	Fe		2Dyp			-2e				$Fe(Dyp)_2^{+2}$
-0.58	Fe		Ox^{-2}			-2e				$Fe(Ox)$
-0.57	Fe		Dge^-			-2e				$Fe(Dge)^+$
-0.57	Fe		En			-2e				$Fe(En)^{+2}$
-0.56	Fe		2Deta			-2e				$Fe(Deta)_2^{+2}$
-0.56	Fe		Dyp			-2e				$Fe(Dyp)^{+2}$
-0.56	Fe		OH^-			-2e				$FeOH^+$
-0.56		OH^-	$Fe(OH)_2$			-e				$Fe(OH)_3$
-0.56	Fe		$Sald^-$			-2e				$Fe(Sald)^+$
-0.54	Fe		2En			-2e				$Fe(En)_2^{+2}$
-0.54	Fe		$2Sald^-$			-2e				$Fe(Sald)_2$
-0.53	Fe		$2Dge^-$			-2e				$Fe(Dge)_2$
-0.52	Fe		3En			-2e				$Fe(En)_3^{+2}$
-0.52	Fe		Mal^{-2}			-2e				$Fe(Mal)$
-0.51	Fe		SO_4^{-2}			-2e				$FeSO_4$
-0.50	Fe		$S_2O_3^{-2}$			-2e				FeS_2O_3
-0.48	Fe		NH_3			-2e				$Fe(NH_3)^{+2}$
-0.46	Fe		$2NH_3$			-2e				$Fe(NH_3)_2^{+2}$
-0.46	Fe		$3Ox^{-2}$			-2e				$Fe(Ox)_3^{-4}$

EMF	Elem	Anion	Compound or anion	Cation	H_2O	e =	Cation	Anion	Complex or compound	Compound or element
-0.44	Fe					-2e	Fe^{+2}			
-0.17	Fe		$Edta^{-4}$			-3e			$Fe(Edta)^-$	
-0.04	Fe					-3e	Fe^{+3}			
+0.01	Fe		Sal^{-2}			-3e			$Fe(Sal)^+$	
+0.10	Fe		OH^-			-3e			$FeOH^{+2}$	
+0.10	Fe		$2Sal^{-2}$			-3e			$Fe(Sal)_2^-$	
+0.12	Fe		$2OH^-$			-3e			$Fe(OH)_2^+$	
+0.14	Fe		$AcAc^-$			-3e			$Fe(AcAc)^{+2}$	
+0.14	Fe		Ox^{-2}			-3e			$Fe(Ox)^+$	
+0.15	Fe		$2AcAc^-$			-3e			$Fe(AcAc)_2^+$	
+0.17	Fe		$2Nta^{-3}$			-3e			$Fe(Nta)_2^{-3}$	
+0.18	Fe		$3AcAc^-$			-3e			$Fe(AcAc)_3$	
+0.19	Fe		Tf^-			-3e			$Fe(Tf)^{+2}$	
+0.20	Fe		$2Ox^{-2}$			-3e			$Fe(Ox)_2^-$	
+0.22	Fe		$3Sal^{-2}$			-3e			$Fe(Sal)_3^{-3}$	
+0.23	Fe		F^-			-3e			FeF^{+2}	
+0.25	Fe		$2F^-$			-3e			FeF_2^+	
+0.25	Fe		$3Ox^{-2}$			-3e			$Fe(Ox)_3^{-3}$	
+0.27	Fe		SO_4^{-2}			-3e			$FeSO_4^+$	
+0.28	Fe		$3F^-$			-3e			FeF_3	
+0.31	Fe		$2SO_4^{-2}$			-3e			$Fe(SO_4)_2^-$	
+0.32	Fe		Cl^-			-3e			$FeCl^{+2}$	
+0.33	Fe		$2Cl^-$			-3e			$FeCl_2^+$	
+0.34	Fe		Br^-			-3e			$FeBr^{+2}$	
+0.36	Fe		$3Cl^-$			-3e			$FeCl_3$	
+0.36			$(FeCy)^{-4}$			-e		$(FeCy)^{-3}$		
+0.72		$5OH^-$	$Fe(OH)_3$			-3e		FeO_4^{-2}	$4H_2O$	
+0.77				Fe^{+2}		-e	Fe^{+3}			
+2.20				Fe^{+3}	$4H_2O$	-3e	$8H^+$	FeO_4^{-2}		

EMF	Elem	Anion	Compound or anion	Cation	H_2O	e =	Cation	Anion	Complex or compound	Compound or element
-2.10	Fm					-3e	Fm^{+3}			

EMF	Elem	Anion	Compound or anion	Cation	H_2O	e =	Cation	Anion	Complex or compound	Compound or element
-1.22	Ga		$4OH^-$			-3e		$H_2GaO_3^-$	H_2O	
-0.98	Ga		$Data^{-4}$			-3e			$Ga(Data)^-$	
-0.74	Ga		OH^-			-3e			$GaOH^{+2}$	
-0.72	Ga		$AcAc^-$			-3e			$Ga(AcAc)^{+2}$	
-0.69	Ga		$2AcAc^-$			-3e			$Ga(AcAc)_2^+$	
-0.68				Ga^{+2}		-e	Ga^{+3}			
-0.65	Ga		$3AcAc^-$			-3e			$Ga(AcAc)_3$	
-0.63	Ga		F^-			-3e			GaF^{+2}	
-0.55	Ga		Cl^-			-3e			$GaCl^{+2}$	
-0.53	Ga					-3e	Ga^{+3}			
-0.51	Ga		$2Cl^-$			-3e			$GaCl_2^+$	
-0.49	2Ga				$3H_2O$	-6e	$6H^+$		Ga_2O_3	
-0.47	Ga		$3Cl^-$			-3e			$GaCl_3$	
-0.45	Ga		$4Cl^-$			-3e			$GaCl_4^-$	
-0.40	2Ga				H_2O	-2e	$2H^+$		Ga_2O	

EMF	Elem	Anion	Compound or anion	Cation	H_2O	e =	Cation	Anion	Complex or compound	Compound or element
-2.82	Gd		$3OH^-$			-3e			$Gd(OH)_3$	
-2.77	Gd		$Data^{-4}$			-3e			$Gd(Data)^-$	
-2.73	Gd		$Edta^{-4}$			-3e			$Gd(Edta)^-$	
-2.51	Gd		$Oxac^{-2}$			-3e			$Gd(Oxac)^+$	
-2.49	Gd		$2Oxac^{-2}$			-3e			$Gd(Oxac)_2^-$	
-2.47	Gd		SO_4^{-2}			-3e			$GdSO_4^+$	
-2.45	Gd		F^-			-3e			GdF^{+2}	
-2.40	Gd					-3e	Gd^{+3}			
-1.99	2Gd				$3H_2O$	-6e	$6H^I$		Gd_2O_3	

EMF	Elem	Anion	Compound or anion	Cation	H_2O	e =	Cation	Anion	Complex or compound	Compound or element
-1.03	Ge		$5OH^-$			-4e		$HGeO_3^-$	$2H_2O$	
-0.87			GeH_4			-4e	$4H^+$				Ge
-0.15	Ge				$2H_2O$	-4e	$4H^+$		GeO_2	
0.00				Ge^{+2}		-2e	Ge^{+4}			
+0.12	Ge					-4e	Ge^{+4}			
+0.24	Ge					-2e	Ge^{+2}			

EMF	Elem	Anion	Compound or anion	Cation	H_2O	e =	Cation	Anion	Complex or compound	Compound or element
0.00	D_2					-2e	$2D^+$			
0.00	H_2					-2e	$2H^+$			

EMF	Elem	Anion	Compound or anion	Cation	H_2O	e =	Cation	Anion	Complex or compound	Compound or element
-2.50	Hf		$4OH^-$			-4e			$HfO(OH)_2$	H_2O
-1.70	Hf					-4e	Hf^{+4}			
-1.51	Hf				$2H_2O$	-4e	$4H^+$		HfO_2	

EMF	Elem	Anion	Compound or anion	Cation	H_2O	e =	Cation	Anion	Complex or compound	Compound or element
-2.28			HgH			-e	H^+			Hg
-0.69	Hg		S^{-2}			-2e			HgS	
-0.43	6Hg		$2(CoCy)^{-3}$			-6e			$(Hg_2)_3(CoCy)_2$	
-0.37	Hg		$4CN^-$			-2e			$Hg(CN)_4^{-2}$	
-0.26	2Hg		$2N_3^-$			-2e			$Hg_2(N_3)_2$	
-0.04	2Hg		$2I^-$			-2e			Hg_2I_2	
-0.04	Hg		$4I^-$			-2e			HgI_4^{-2}	
+0.03	Hg		$2S_2O_3^{-2}$			-2e			$Hg(S_2O_3)_2^{-2}$	
+0.03	Hg		$4SeCN^-$			-2e			$Hg(SeCN)_4^{-2}$	
+0.10	Hg		$2OH^-$			-2e			HgO (r)	H_2O
+0.14	2Hg		$2Br^-$			-2e			Hg_2Br_2	
+0.22	Hg		$4Br^-$			-2e			$HgBr_4^{-2}$	
+0.27	2Hg		$2Cl^-$			-2e			Hg_2Cl_2	
+0.27	Hg		$4SCN^-$			-2e			$Hg(SCN)_4^{-2}$	
+0.39	2Hg		$2IO_3^-$			-2e			$Hg(IO_3)_2$	
+0.51	2Hg		$2Ac^-$			-2e			$Hg_2(Ac)_2$	
+0.62	2Hg		SO_4^{-2}			-2e			Hg_2SO_4	
+0.64	2Hg		HPO_4^{-2}			-2e			Hg_2HPO_4	
+0.79	2Hg					-2e	Hg_2^{+2}			
+0.88	Hg	F^-				-2e			HgF^+	
+0.92				Hg_2^{+2}		-2e	$2Hg^{+2}$			
+0.93	Hg				H_2O	-2e	$2H^+$		HgO	
+1.03	Hg				$2H_2O$	-2e	$2H^+$		$Hg(OH)_2$	

EMF	Elem	Anion	Compound or anion	Cation	H_2O	e =	Cation	Anion	Complex or compound	Compound or element
-2.77	Ho		$3OH^-$			-3e			$Ho(OH)_3$	
-2.67	Ho		$Edta^{-4}$			-3e			$Ho(Edta)^-$	
-2.39	Ho		SO_4^{-2}			-3e			$HoSO_4^+$	
-2.32	Ho					-3e	Ho^{+3}			
-1.94	2Ho				$3H_2O$	-6e	$6H^+$		Ho_2O_3	

EMF	Elem	Anion	Compound or anion	Cation	H_2O	e =	Cation	Anion	Complex or compound	Compound or element
-0.59	H		IO^-			-e			HIO	
-0.09	H		IO_4^-			-e			HIO_4	
-0.05	H		IO_3^-			-e			HIO_3	
+0.26		$6OH^-$	I^-			-6e		IO_3^-		$3H_2O$
+0.49		$2OH^-$	I^-			-2e		IO^-	H_2O	
+0.54			$2I^-$			-2e				I_2 (c)
+0.54			$3I^-$			-2e		I_3^-		
+0.65	H		I^-			-e			HI	
+0.70		$3OH^-$	IO_3^-			-2e		$H_3IO_6^{-2}$		
+1.06	$\tfrac{1}{2}I_2$		$2Cl^-$			-e		ICl_2^-		
+1.20	$\tfrac{1}{2}I_2$				$3H_2O$	-5e	$6H^+$	IO_3^-		
+1.26	I_2					-2e	$2I^+$			
+1.45	$\tfrac{1}{2}I_2$				H_2O	-e	H^+		HIO	
+1.60			IO_3^-		$3H_2O$	-2e	H^+		H_5IO_6	

EMF	Elem	Anion	Compound or anion	Cation	H_2O	e =	Cation	Anion	Complex or compound	Compound or element
-1.95			InH			-e	H^+			In
-1.00	In		$3OH^-$			-3e			$In(OH)_3$	
-0.54	In		OH^-			-3e			$InOH^{+2}$	
-0.50	In		$AcAc^-$			-3e			$In(AcAc)^{+2}$	
-0.49				In^{+2}		-e	In^{+3}			
-0.48	In		$2AcAc^-$			-3e			$In(AcAc)_2^+$	
-0.44				In^+		-2e	In^{+3}			
-0.41	In		Ac^-			-3e			$In(Ac)^{+2}$	
-0.41	In		F^-			-3e			InF^{+2}	
-0.40				In^+		-e	In^{+2}			
-0.39	In		$2Ac^-$			-3e			$In(Ac)_2^+$	
-0.39	In		CNS^-			-3e			$InCNS^{+2}$	
-0.39	In		$2F^-$			-3e			InF_2^+	
-0.39	In		$3F^-$			-3e			InF_3	
-0.38	In		$3Ac^-$			-3e			$In(Ac)_3$	
-0.38	In		SO_4^{-2}			-3e			$InSO_4^+$	
-0.37	In		Cl^-			-3e			$InCl^{+2}$	
-0.37	In		$3CNS^-$			-3e			$In(CNS)_3$	
-0.36	In		$4Ac^-$			-3e			$In(Ac)_4^-$	
-0.36	In		$6Ac^-$			-3e			$In(Ac)_6^{-3}$	
-0.36	In		Br^-			-3e			$InBr^{+2}$	
-0.36	In		$2Cl^-$			-3e			$InCl_2^+$	
-0.36	In		$3Cl^-$			-3e			$InCl_3$	
-0.36	In		$4F^-$			-3e			InF_4^-	
-0.35	In		$2Br^-$			-3e			$InBr_2^+$	
-0.35	In		$3Br^-$			-3e			$InBr_3$	
-0.35	In		$2CNS^-$			-3e			$In(CNS)_2^+$	
-0.35	In		I^-			-3e			InI^{+2}	
-0.35	In		$3SO_4^{-2}$			-3e			$In(SO_4)_3^{-3}$	
-0.34	In					-3e	In^{+3}			
-0.34	In		$5Ac^-$			-3e			$In(Ac)_5^{-2}$	
-0.34	In		$2SO_4^{-2}$			-3e			$In(SO_4)_2^-$	

EMF	Elem	Anion	Compound or anion	Cation	H_2O	e =	Cation	Anion	Complex or compound	Compound or element
-0.19	2In				$3H_2O$	-6e	$6H^+$		In_2O_3	
-0.17	In				$3H_2O$	-3e	$3H^+$		$In(OH)_3$	
-0.14	In					-e	In^+			

EMF	Elem	Anion	Compound or anion	Cation	H_2O	e =	Cation	Anion	Complex or compound	Compound or element
+0.10	2Ir		$6OH^-$			-6e			Ir_2O_3	$3H_2O$
+0.77	Ir		$6Cl^-$			-3e			$IrCl_6^{-3}$	
+0.87	Ir				H_2O	-2e	$2H^+$		IrO	
+0.93	Ir				$2H_2O$	-4e	$4H^+$		IrO_2	
+0.93	2Ir				$3H_2O$	-6e	$6H^+$		Ir_2O_3	
+0.99			$IrBr_6^{-4}$			-e		$IrBr_6^{-3}$		
+1.02			$IrCl_6^{-3}$			-e		$IrCl_6^{-2}$		
+1.16	Ir					-3e	Ir^{+3}			

EMF	Elem	Anion	Compound or anion	Cation	H_2O	e =	Cation	Anion	Complex or compound	Compound or element
-3.07	K		$(FeCy)^{-4}$			-e			$K(FeCy)^{-3}$	
-3.01	K		$(FeCy)^{-3}$			-e			$K(FeCy)^{-2}$	
-3.01	K		$(CoCy)^{-3}$			-e			$K(CoCy)^{-2}$	
-2.98	K		SO_4^{-2}			-e			KSO_4^{-}	
-2.97	K		ReO_4^{-}			-e			$KReO_4$	
-2.97	K		$S_2O_3^{-2}$			-e			$KS_2O_3^{-}$	
-2.93	K		ClO_3^{-}			-e			$KClO_3$	
-2.93	K					-e	K^+			
-2.92	K		NO_3^{-}			-e			KNO_3	
-2.90	K		IO_3^{-}			-e			KIO_3	
-2.90	K		ClO_4^{-}			-e			$KClO_4$	
-2.90	K		BrO_3^{-}			-e			$KBrO_3$	
-1.27			KH			-2e	K^+			H^+
+0.39			KH			-e	H^+			K

EMF	Elem	Anion	Compound or anion	Cation	H_2O	e =	Cation	Anion	Complex or compound	Compound or element
-2.90	La		$3OH^-$			-3e			$La(OH)_3$	
-2.84	La		$Data^{-4}$			-3e			$La(Data)^-$	
-2.81	La		$Edta^{-4}$			-3e			$La(Edta)^-$	
-2.78	2La		$3S^{-2}$			-6e			La_2S_3	
-2.72	La		$2Amac^{-3}$			-3e			$La(Amac)_2^{-3}$	
-2.72	La		Nta^{-3}			-3e			$La(Nta)$	
-2.69	La		$Oxin^-$			-3e			$La(Oxin)^{+2}$	
-2.67	La		$2Nta^{-3}$			-3e			$La(Nta)_2^{-3}$	
-2.67	La		$2Oxin^-$			-3e			$La(Oxin)_2^+$	
-2.65	La		$P_4O_{12}^{-4}$			-3e			$LaP_4O_{12}^-$	
-2.63	La		$P_3O_9^{-3}$			-3e			LaP_3O_9	
-2.62	La		$AcAc^-$			-3e			$La(AcAc)^{+2}$	
-2.62	La		OH^-			-3e			$La(OH)^{+2}$	
-2.62	La		$Oxac^{-2}$			-3e			$La(Oxac)^+$	
-2.60	La		$(CoCy)^{-3}$			-3e			$La(CoCy)$	
-2.60	La		$(FeCy)^{-3}$			-3e			$La(FeCy)$	
-2.59	La		SO_4^{-2}			-3e			$LaSO_4^+$	
-2.59	La		$2AcAc^-$			-3e			$La(AcAc)_2^+$	
-2.58	La		$3AcAc^-$			-3e			$La(AcAc)_3$	
-2.57	La		F^-			-3e			LaF^{+2}	
-2.52	La					-3e	La^{+3}			
-1.86	2La				$3H_2O$	-6e	$6H^+$			La_2O_3

EMF	Elem	Anion	Compound or anion	Cation	H_2O	e =	Cation	Anion	Complex or compound	Compound or element
-3.34	Li		$Amac^{-3}$			-e			$LiAmac^{-2}$	
-3.21	Li		Nta^{-3}			-e			$LiNta^{-2}$	
-3.19	Li		$Edta^{-4}$			-e			$LiEdta^{-3}$	
-3.15	Li		Saa^{-3}			-e			$LiSaa^{-2}$	
-3.05	Li					-e	Li^+			
-3.05	Li		Lac^{-2}			-e			$Li(Lac)^-$	
-3.03	Li		OH^-			-e			$LiOH$	
-3.03	Li		Gly^{-2}			-e			$Li(Gly)^-$	
-1.16			LiH			-2e	Li^+			H^+
+0.73			LiH			-e	H^+			Li

EMF	Elem	Anion	Compound or anion	Cation	H_2O	e =	Cation	Anion	Complex or compound	Compound or element
-2.72	Lu		$3OH^-$			-3e			$Lu(OH)_3$	
-2.68	Lu		$Data^{-4}$			-3e			$Lu(Data)^-$	
-2.64	Lu		$Edta^{-4}$			-3e			$Lu(Edta)^-$	
-2.38	Lu		$Oxac^{-2}$			-3e			$Lu(Oxac)^+$	
-2.35	Lu		$2Oxac^{-2}$			-3e			$Lu(Oxac)_2^-$	
-2.26	Lu					-3e	Lu^{+3}			
-1.89	2Lu				$3H_2O$	-6e	$6H^+$			Lu_2O_3

EMF	Elem	Anion	Compound or anion	Cation	H_2O	e =	Cation	Anion	Complex or compound	Compound or element
-2.69	Mg		$2OH^-$			-2e				$Mg(OH)_2$
-2.60	Mg		$Amac^{-3}$			-2e				$Mg(Amac)^-$
-2.60	Mg		$Edta^{-4}$			-2e				$Mg(Edta)^{-2}$
-2.56	Mg		Ers^{-3}			-2e				$Mg(Ers)^-$
-2.56	Mg		Esb^{-3}			-2e				$Mg(Esb)^-$
-2.56	Mg		Nta^{-3}			-2e				$Mg(Nta)^-$
-2.55	Mg		Esa^{-3}			-2e				$Mg(Esa)^-$
-2.55	Mg		Est^{-3}			-2e				$Mg(Est)^-$
-2.53	Mg		$Oxin^-$			-2e				$Mg(Oxin)^+$
-2.52	Mg		$Ist(a)^-$			-2e				$Mg(Ist(a))^+$
-2.52	Mg		$Ist(b)^-$			-2e				$Mg(Ist(b))^+$
-2.52	Mg		$Met(a)^-$			-2e				$Mg(Met(a))^+$
-2.52	Mg		$Met(b)^-$			-2e				$Mg(Met(b))^+$
-2.52	Mg		$Tmta^{-4}$			-2e				$Mg(Tmta)^{-2}$
-2.51	Mg		$P_2O_7^{-4}$			-2e				$MgP_2O_7^{-2}$
-2.50	Mg		$2Oxin^-$			-2e				$Mg(Oxin)_2$
-2.50	Mg		$Trop^-$			-2e				$Mg(Trop)^+$
-2.49	Mg		$2Ist(a)^-$			-2e				$Mg(Ist(a))_2$
-2.49	Mg		$MeOxin^-$			-2e				$Mg(MeOxin)^+$
-2.49	Mg		$Ndap^{-3}$			-2e				$Mg(Ndap)^-$
-2.49	Mg		$mOxin^-$			-2e				$Mg(mOxin)^+$
-2.49	Mg		$P_4O_{12}^{-4}$			-2e				$MgP_4O_{12}^{-2}$
-2.48	Mg		$2Ist(b)^-$			-2e				$Mg(Ist(b))_2$
-2.48	Mg		$(FeCy)^{-4}$			-2e				$Mg(FeCy)^{-2}$
-2.48	Mg		$2Met(a)^-$			-2e				$Mg(Met(a))_2$
-2.48	Mg		$2Met(b)^-$			-2e				$Mg(Met(b))_2$
-2.48	Mg		$meOxin^-$			-2e				$Mg(meOxin)^+$
-2.47	Mg		$2MeOxin^-$			-2e				$Mg(MeOxin)_2$
-2.47	Mg		$2mOxin^-$			-2e				$Mg(mOxin)_2$
-2.47	Mg		$2Trop^-$			-2e				$Mg(Trop)_2$
-2.46	Mg		Has^-			-2e				$Mg(Has)^+$
-2.46	Mg		$2meOxin^-$			-2e				$Mg(meOxin)_2$

EMF	Elem	Anion	Compound or anion	Cation	H_2O	e =	Cation	Anion	Complex or compound	Compound or element
-2.45	Mg		$AcAc^-$			-2e			$Mg(AcAc)^+$	
-2.45	Mg		$DmHas^-$			-2e			$Mg(DmHas)^+$	
-2.45	Mg		$Imda^{-2}$			-2e			$Mg(Imda)$	
-2.45	Mg		$Mcin^-$			-2e			$Mg(Mcin)^+$	
-2.45	Mg		$Ndpa^{-3}$			-2e			$Mg(Ndpa)^-$	
-2.45	Mg		Oxd^-			-2e			$Mg(Oxd)^+$	
-2.44	Mg		Gl^-			-2e			$Mg(Gl)^+$	
-2.44	Mg		$Himda^{-2}$			-2e			$Mg(Himda)$	
-2.44	Mg		Hox^-			-2e			$Mg(Hox)^+$	
-2.44	Mg		PO_4^{-3}			-2e			$MgPO_4^-$	
-2.44	Mg		Ox^{-2}			-2e			$Mg(Ox)$	
-2.44	Mg		$P_3O_9^{-3}$			-2e			$MgP_3O_9^-$	
-2.43	Mg		$2Amac^{-3}$			-2e			$Mg(Amac)_2^{-4}$	
-2.43	Mg		Cin^-			-2e			$Mg(Cin)^+$	
-2.43	Mg		Cit^{-3}			-2e			$Mg(Cit)^-$	
-2.43	Mg		$2DmHas^-$			-2e			$Mg(DmHas)_2$	
-2.43	Mg		$2Has^-$			-2e			$Mg(Has)_2$	
-2.43	Mg		$2Hox^-$			-2e			$Mg(Hox)_2$	
-2.43	Mg		$3Ist(b)^-$			-2e			$Mg(Ist(b))_3^-$	
-2.43	Mg		$3Met(b)^-$			-2e			$Mg(Met(b))_3^-$	
-2.43	Mg		$2Nta^{-3}$			-2e			$Mg(Nta)_2^{-4}$	
-2.43	Mg		$2Oxd^-$			-2e			$Mg(Oxd)_2$	
-2.42	Mg		$2AcAc^-$			-2e			$Mg(AcAc)_2$	
-2.42	Mg		Mal^{-2}			-2e			$Mg(Mal)$	
-2.42	Mg		$2Mcin^-$			-2e			$Mg(Mcin)_2$	
-2.42	Mg		$3Met(a)^-$			-2e			$Mg(Met(a))_3^-$	
-2.42	Mg		OH^-			-2e			$MgOH^+$	
-2.42	Mg		Saa^{-3}			-2e			$Mg(Saa)^-$	
-2.41	Mg		$Aspa^{-2}$			-2e			$Mg(Aspa)$	
-2.41	Mg		HPO_4^{-2}			-2e			$MgHPO_4$	
-2.41	Mg		SO_4^{-2}			-2e			$MgSO_4$	
-2.40	Mg		$Alan^-$			-2e			$Mg(Alan)^+$	

EMF	Elem	Anion	Compound or anion	Cation	H_2O	e =	Cation	Anion	Complex or compound	Compound or element
-2.40	Mg		$2Cin^-$			-2e			$Mg(Cin)_2$	
-2.39	Mg		Ap^{-2}			-2e			$Mg(Ap)$	
-2.39	Mg		$S_2O_3^{-2}$			-2e			MgS_2O_3	
-2.38	Mg		F^-			-2e			MgF^+	
-2.38	Mg		Ph			-2e			$Mg(Ph)^{+2}$	
-2.38	Mg		Suc^{-2}			-2e			$Mg(Suc)$	
-2.38	Mg		$Tart^{-2}$			-2e			$Mg(Tart)$	
-2.37	Mg		Dge^-			-2e			$Mg(Dge)^+$	
-2.37	Mg		$Glac^-$			-2e			$Mg(Glac)^+$	
-2.37	Mg		$Glgl^-$			-2e			$Mg(Glgl)^+$	
-2.37	Mg		Gly^-			-2e			$Mg(Gly)^+$	
-2.37	Mg		Lac^-			-2e			$Mg(Lac)^+$	
-2.36	Mg					-2e	Mg^{+2}			
-2.36	Mg		Ac^-			-2e			$Mg(Ac)^+$	
-2.36	Mg		Bu^-			-2e			$Mg(Bu)^+$	
-2.36	Mg		Glu^-			-2e			$Mg(Glu)^+$	
-2.36	Mg		IO_3^-			-2e			$MgIO_3^+$	
-2.36	Mg		Pr^-			-2e			$Mg(Pr)^+$	
-2.35	Mg		Dyp			-2e			$Mg(Dyp)^{+2}$	
-2.35	Mg		NH_3			-2e			$Mg(NH_3)^{+2}$	
-2.34	Mg		$2NH_3$			-2e			$Mg(NH_3)_2^{+2}$	
-2.34	Mg		NO_3^-			-2e			$MgNO_3^+$	
-2.33	Mg		Nac^-			-2e			$Mg(Nac)^+$	
-2.33	Mg		$3NH_3$			-2e			$Mg(NH_3)_3^{+2}$	
-2.32	Mg		$4NH_3$			-2e			$Mg(NH_3)_4^{+2}$	
-2.31	Mg		$5NH_3$			-2e			$Mg(NH_3)_5^{+2}$	
-2.30	Mg		$6NH_3$			-2e			$Mg(NH_3)_6^{+2}$	

EMF	Elem	Anion	Compound or anion	Cation	H$_2$O	e =	Cation	Anion	Complex or compound	Compound or element
-1.55	Mn		Data^{-4}			-2e			Mn(Data)$^{-2}$	
-1.55	Mn		2OH$^-$			-2e			Mn(OH)$_2$	
-1.50	Mn		CO$_3^{-2}$			-2e			MnCO$_3$ (c)	
-1.48	Mn		CO$_3^{-2}$			-2e			MnCO$_3$ (ppt)	
-1.45	Mn		Edta^{-4}			-2e			Mn(Edta)$^{-2}$	
-1.37	Mn		Hed^{-3}			-2e			Mn(Hed)$^-$	
-1.29	Mn		Oxin$^-$			-2e			Mn(Oxin)$^+$	
-1.28	Mn		Oxd$^-$			-2e			Mn(Oxd)$^+$	
-1.27	Mn		Nta^{-3}			-2e			Mn(Nta)$^-$	
-1.26	Mn		2Oxin$^-$			-2e			Mn(Oxin)$_2$	
-1.25	Mn		2Oxd$^-$			-2e			Mn(Oxd)$_2$	
-1.22	Mn		Himda^{-2}			-2e			Mn(Himda)	
-1.22	Mn		Tate			-2e			Mn(Tate)$^{+2}$	
-1.21	Mn		P$_4$O$_{12}^{-4}$			-2e			MnP$_4$O$_{12}^{-2}$	
-1.19	Mn		Teta			-2e			Mn(Teta)$^{+2}$	
-1.18	Mn					-2e	Mn^{+2}			
-1.18	Mn		AcAc$^-$			-2e			Mn(AcAc)$^+$	
-1.17	Mn		2Amac^{-3}			-2e			Mn(Amac)$_2^{-4}$	
-1.17	Mn		Aspa^{-2}			-2e			Mn(Aspa)	
-1.17	Mn		Deta			-2e			Mn(Deta)$^{+2}$	
-1.17	Mn		2Himda^{-2}			-2e			Mn(Himda)$_2^{-2}$	
-1.17	Mn		Ox^{-2}			-2e			Mn(Ox)	
-1.16	Mn		2Nta^{-3}			-2e			Mn(Nta)$_2^{-4}$	
-1.16	Mn		P$_3$O$_9^{-3}$			-2e			MnP$_3$O$_9^-$	
-1.16	Mn		Sald$^-$			-2e			Mn(Sald)$^+$	
-1.15	Mn		Gl$^-$			-2e			Mn(Gl)$^+$	
-1.15	Mn		Mal^{-2}			-2e			Mn(Mal)	
-1.15	Mn		OH$^-$			-2e			MnOH$^+$	
-1.14	Mn		2AcAc$^-$			-2e			Mn(AcAc)$_2$	
-1.14	Mn		Alan$^-$			-2e			Mn(Alan)$^+$	
-1.14	Mn		2Alan$^-$			-2e			Mn(Alan)$_2$	
-1.14	Mn		Dge$^-$			-2e			Mn(Dge)$^+$	

EMF	Elem	Anion	Compound or anion	Cation	H$_2$O	e =	Cation	Anion	Complex or compound	Compound or element
-1.14	Mn		2Sald$^-$			-2e			Mn(Sald)$_2$	
-1.13	Mn		2Deta			-2e			Mn(Deta)$_2^{+2}$	
-1.13	Mn		En			-2e			Mn(En)$^{+2}$	
-1.12	Mn		2Dge$^-$			-2e			Mn(Dge)$_2$	
-1.12	Mn		Dyp			-2e			Mn(Dyp)$^{+2}$	
-1.12	Mn		SO$_4^{-2}$			-2e			MnSO$_4$	
-1.11	Mn		2En			-2e			Mn(En)$_2^{+2}$	
-1.11	Mn		Glgl$^-$			-2e			Mn(Glgl)$^+$	
-1.11	Mn		S$_2$O$_3^{-2}$			-2e			MnS$_2$O$_3$	
-1.09	Mn		Ac$^-$			-2e			Mn(Ac)$^+$	
-1.08	Mn		3En			-2e			Mn(En)$_3^{+2}$	
-1.07	Mn		NH$_3$			-2e			Mn(NH$_3$)$^{+2}$	
-1.06	Mn		2NH$_3$			-2e			Mn(NH$_3$)$_2^{+2}$	
-0.05		2OH$^-$	Mn(OH)$_2$			-2e			MnO$_2$	2H$_2$O
+0.13	Mn		Ox^{-2}			-3e			Mn(Ox)$^+$	
+0.14	H		MnO$_4^-$			-e			HMnO$_4$	
+0.15		OH$^-$	Mn(OH)$_2$			-e			Mn(OH)$_3$	
+0.20	Mn		2Ox^{-2}			-3e			Mn(Ox)$_2^=$	
+0.22	Mn		F$^-$			-3e			MnF^{+2}	
+0.27	Mn		3Ox^{-2}			-3e			Mn(Ox)$_3^{-3}$	
+0.31	Mn		Cl$^-$			-3e			MnCl^{+2}	
+0.56		MnO$_4^{-2}$				-e		MnO$_4^-$		
+0.59	4OH$^-$	MnO$_2$				-3e		MnO$_4^-$	2H$_2$O	
+0.60	4OH$^-$	MnO$_2$				-2e		MnO$_4^{-2}$	2H$_2$O	
+1.23				Mn^{+2}	2H$_2$O	-2e	4H$^+$		MnO$_2$	
+1.51				Mn^{+2}		-e	Mn^{+3}			
+1.51				Mn^{+2}	4H$_2$O	-5e	8H$^+$	MnO$_4^-$		
+1.70			MnO$_2$		2H$_2$O	-3e	4H$^+$	MnO$_4^-$		

EMF	Elem	Anion	Compound or anion	Cation	H_2O	e =	Cation	Anion	Complex or compound	Compound or element
-1.05	Mo		$8OH^-$			-6e		MoO_4^{-2}	$4H_2O$	
-0.20	Mo					-3e	Mo^{+3}			
-0.07	Mo				$2H_2O$	-4e	$4H^+$		MoO_2	
+0.15	Mo				$4H_2O$	-6e	$8H^+$	MoO_4^{-2}		
+0.32			MoO_2		H_2O	-2e	$2H^+$		MoO_3	
+0.32				Mo^{+3}	$3H_2O$	-3e	$6H^+$		MoO_3	

EMF	Elem	Anion	Compound or anion	Cation	H_2O	e =	Cation	Anion	Complex or compound	Compound or element
-3.40			HN_3 (g)			-e	H^+			3/2 N_2
-3.09			HN_3 (aq)			-e	H^+			3/2 N_2
-1.95	H	NH_2^-				-e			NH_3	
-0.97		CN^-	$2OH^-$			-2e		CNO^-	H_2O	
-0.54	H		CN^-			-e			HCN	
-0.39	H		$O_2N \cdot NH^-$			-e			$O_2N \cdot NH_2$	
-0.33			$\frac{1}{2}C_2N_2$		H_2O	-e	H^+		HCNO	
-0.23				$N_2H_5^+$		-4e	$5H^+$			N_2
-0.20	H		NO_2^-			-e			HNO_2	
-0.01	H		$UREA^-$			-e			UREA	
-0.01	2H		$UREA^{-2}$			-2e			UREA	
0.00	3H		$UREA^{-3}$			-3e			UREA	
+0.01		$2OH^-$	NO_2^-			-2e		NO_3^-		H_2O
+0.06			$2NH_3$			-6e	$6H^+$			N_2
+0.10	H		NO_3^-			-e			HNO_3	
+0.11		$2OH^-$	$2NH_4OH$			-2e			N_2H_4	$4H_2O$
+0.37			HCN (aq)			-e	H^+		$\frac{1}{2}C_2N_2$ (g)	
+0.39			$2NH_3OH^+$			-4e	$6H^+$		$H_2N_2O_2$	
+0.70				$3NH_4^+$		-8e	$11H^+$		HN_3 (aq)	
+0.71			$H_2N_2O_2$			-2e	$2H^+$		2NO	
+0.73		$2OH^-$	N_2H_4			-2e			$2NH_2OH$	
+0.77			$2CNS^-$			-2e			$(CNS)_2$	
+0.80			N_2O_4 (g)		$2H_2O$	-2e	$4H^+$	$2NO_3^-$		
+0.86			$H_2N_2O_2$		$2H_2O$	-4e	$4H^+$		$2HNO_2$	
+0.94			HNO_2		H_2O	-2e	$3H^+$	NO_3^-		
+0.96			NO		$2H_2O$	-3e	$4H^+$	NO_3^-		
+1.00			NO		H_2O	-e	H^+		HNO_2	
+1.03			2NO		$2H_2O$	-4e	$4H^+$		N_2O_4	
+1.07			$2HNO_2$			-2e	$2H^+$		N_2O_4	
+1.09			HNO_2			-e	H^+		NO_2	
+1.28				$2NH_4^+$		-2e	$N_2H_5^+$			$3H^+$
+1.29			N_2O (g)		$3H_2O$	-4e	$4H^+$		$2HNO_2$(aq)	

EMF	Elem	Anion	Compound or anion	Cation	H_2O	e =	Cation	Anion	Complex or compound	Compound or element
+1.35				NH_4^+	H_2O	-2e	NH_3OH^+			$2H^+$
+1.42				$N_2H_5^+$	$2H_2O$	-2e	H^+		$2NH_3OH^+$	
+1.45	N_2				$4H_2O$	-6e	$6H^+$		$2HNO_2$	
+1.96	N_2			NH_4^+		-2e	$3H^+$		HN_3	
+2.65	N_2				$2H_2O$	-2e	$2H^+$		$H_2N_2O_2$	

EMF	Elem	Anion	Compound or anion	Cation	H_2O	e =	Cation	Anion	Complex or compound	Compound or element
-2.91	Na		$Amac^{-3}$			-e			$Na(Amac)^{-2}$	
-2.85	Na		$P_2O_7^{-4}$			-e			$Na(P_2O_7)^{-3}$	
-2.84	Na		Nta^{-3}			-e			$Na(Nta)^{-2}$	
-2.81	Na		$Edta^{-4}$			-e			$Na(Edta)^{-3}$	
-2.78	Na		$P_3O_9^{-3}$			-e			$NaP_3O_9^{-2}$	
-2.77	Na		Saa^{-3}			-e			$Na(Saa)^{-2}$	
-2.76	Na		$P_4O_{12}^{-4}$			-e			$NaP_4O_{12}^{-3}$	
-2.75	Na		$S_2O_3^{-2}$			-e			$NaS_2O_3^{-}$	
-2.75	Na		SO_4^{-2}			-e			$NaSO_4^{-}$	
-2.71	Na					-e	Na^+			
-2.69	Na		BrO_3^{-}			-e			$NaBrO_3$	
-2.68	Na		OH^{-}			-e			$NaOH$	
-1.16			NaH			-2e	Na^+			H^+
+0.39			NaH			-e	H^+			Na

EMF	Elem	Anion	Compound or anion	Cation	H_2O	e =	Cation	Anion	Complex or compound	Compound or element
-1.10	Nb					-3e	Nb^{+3}			
-0.73	Nb				H_2O	-2e	$2H^+$		NbO	
-0.64	2Nb				$5H_2O$	-10e	$10H^+$		Nb_2O_5	
-0.63			NbO		H_2O	-2e	$2H^+$		NbO_2	
-0.29			$2NbO_2$		H_2O	-2e	$2H^+$		Nb_2O_5	

EMF	Elem	Anion	Compound or anion	Cation	H_2O	e =	Cation	Anion	Complex or compound	Compound or element
-2.84	Nd		$3OH^-$			-3e			$Nd(OH)_3$	
-2.78	Nd		$Data^{-4}$			-3e			$Nd(Data)^-$	
-2.75	Nd		$Edta^{-4}$			-3e			$Nd(Edta)^-$	
-2.57	Nd		Ox^{-2}			-3e			$Nd(Ox)^+$	
-2.54	Nd		$AcAc^-$			-3e			$Nd(AcAc)^{+2}$	
-2.51	Nd		$2Ox^{-2}$			-3e			$Nd(Ox)_2^-$	
-2.50	Nd		SO_4^{-2}			-3e			$NdSO_4^+$	
-2.49	Nd		$3AcAc^-$			-3e			$Nd(AcAc)_3$	
-2.43	Nd					-3e	Nd^{+3}			
-1.81	2Nd				$3H_2O$	-6e	$6H^+$		Nd_2O_3	
-0.51	Nd		$2AcAc^-$			-3e			$Nd(AcAc)_2^+$	

EMF	Elem	Anion	Compound or anion	Cation	H_2O	e =	Cation	Anion	Complex or compound	Compound or element
-1.04	Ni		S^{-2}			-2e			NiS (gm)	
-0.83	Ni		S^{-2}			-2e			NiS (a)	
-0.80	Ni		$Edta^{-4}$			-2e			$Ni(Edta)^{-2}$	
-0.75	Ni		Hed^{-3}			-2e			$Ni(Hed)^{-}$	
-0.72	Ni		$2OH^{-}$			-2e			$Ni(OH)_2$	
-0.69	Ni		Tate			-2e			$Ni(Tate)^{+2}$	
-0.66	Ni		Teta			-2e			$Ni(Teta)^{+2}$	
-0.59	Ni		$Ndap^{-3}$			-2e			$Ni(Ndap)^{-}$	
-0.59	Ni		$Oxin^{-}$			-2e			$Ni(Oxin)^{+}$	
-0.58	Ni		Nta^{-3}			-2e			$Ni(Nta)^{-}$	
-0.57	Ni		Deta			-2e			$Ni(Deta)^{+2}$	
-0.54	Ni		$2Oxin^{-}$			-2e			$Ni(Oxin)_2$	
-0.53	Ni		$Himda^{-2}$			-2e			$Ni(Himda)$	
-0.53	Ni		Ptn			-2e			$Ni(Ptn)^{+2}$	
-0.52	Ni		$Ndpa^{-3}$			-2e			$Ni(Ndpa)^{-}$	
-0.50	Ni		$Ist(a)^{-}$			-2e			$Ni(Ist(a))^{+}$	
-0.50	Ni		$Ist(b)^{-}$			-2e			$Ni(Ist(b))^{+}$	
-0.50	Ni		$Mcin^{-}$			-2e			$Ni(Mcin)^{+}$	
-0.50	Ni		$Met(a)^{-}$			-2e			$Ni(Met(a))^{+}$	
-0.50	Ni		$Met(b)^{-}$			-2e			$Ni(Met(b))^{+}$	
-0.50	Ni		Oxd^{-}			-2e			$Ni(Oxd)^{+}$	
-0.49	Ni		Cin^{-}			-2e			$Ni(Cin)^{+}$	
-0.49	Ni		2Deta			-2e			$Ni(Deta)_2^{+2}$	
-0.49	Ni		$Imda^{-2}$			-2e			$Ni(Imda)$	
-0.49	Ni		$2Mcin^{-}$			-2e			$Ni(Mcin)_2$	
-0.49	Ni		$2Oxd^{-}$			-2e			$Ni(Oxd)_2$	
-0.48	Ni		$6NH_3(aq)$			-2e			$Ni(NH_3)_6^{+2}$	
-0.48	Ni		$DmHas^{-}$			-2e			$Ni(DmHas)^{+}$	
-0.48	Ni		En			-2e			$Ni(En)^{+2}$	
-0.48	Ni		Hox^{-}			-2e			$Ni(Hox)^{+}$	
-0.48	Ni		$Trop^{-}$			-2e			$Ni(Trop)^{+}$	
-0.47	Ni		$Impa^{-2}$			-2e			$Ni(Impa)$	

EMF	Elem	Anion	Compound or anion	Cation	H_2O	e =	Cation	Anion	Complex or compound	Compound or element
-0.47	Ni		Pn			-2e			$Ni(Pn)^{+2}$	
-0.46	Ni		$Aspa^{-2}$			-2e			$Ni(Aspa)$	
-0.46	Ni		$2Cin^-$			-2e			$Ni(Cin)_2$	
-0.46	Ni		Coy^-			-2e			$Ni(Coy)^+$	
-0.46	Ni		$2DmHas^-$			-2e			$Ni(DmHas)_2$	
-0.46	Ni		$2Hox^-$			-2e			$Ni(Hox)_2$	
-0.45	Ni		CO_3^{-2}			-2e			$NiCO_3$	
-0.45	Ni		Af^-			-2e			$Ni(Af)^+$	
-0.45	Ni		$2Ist(a)^-$			-2e			$Ni(Ist(a))_2$	
-0.45	Ni		$2Met(a)^-$			-2e			$Ni(Met(a))_2$	
-0.45	Ni		$2Met(b)^-$			-2e			$Ni(Met(b))_2$	
-0.44	Ni		Dge^-			-2e			$Ni(Dge)^+$	
-0.44	Ni		2En			-2e			$Ni(En)_2^{+2}$	
-0.44	Ni		$2Imda^{-2}$			-2e			$Ni(Imda)_2^{-2}$	
-0.44	Ni		$2Ist(b)^-$			-2e			$Ni(Ist(b))_2$	
-0.44	Ni		2Pn			-2e			$Ni(Pn)_2^{+2}$	
-0.44	Ni		Tmen			-2e			$Ni(Tmen)^{+2}$	
-0.43	Ni		$AcAc^-$			-2e			$Ni(AcAc)^+$	
-0.43	Ni		$Alan^-$			-2e			$Ni(Alan)^+$	
-0.43	Ni		$Imdp^{-2}$			-2e			$Ni(Imdp)$	
-0.43	Ni		$2Trop^-$			-2e			$Ni(Trop)_2$	
-0.42	Ni		Gl^-			-2e			$Ni(Gl)^+$	
-0.42	Ni		Ntp^{-3}			-2e			$Ni(Ntp)^-$	
-0.42	Ni		$P_2O_7^{-4}$			-2e			$NiP_2O_7^{-2}$	
-0.41	Ni		$2Aspa^{-2}$			-2e			$Ni(Aspa)_2^{-2}$	
-0.41	Ni		$2Coy^-$			-2e			$Ni(Coy)_2$	
-0.41	Ni		Ox^{-2}			-2e			$Ni(Ox)$	
-0.40	Ni		$2Himda^{-2}$			-2e			$Ni(Himda)_2^{-2}$	
-0.40	Ni		$2Impa^{-2}$			-2e			$Ni(Impa)_2^{-2}$	
-0.40	Ni		$P_4O_{12}^{-4}$			-2e			$NiP_4O_{12}^{-2}$	
-0.40	Ni		$Sald^-$			-2e			$Ni(Sald)^+$	
-0.39	Ni		$2AcAc^-$			-2e			$Ni(AcAc)_2$	

EMF	Elem	Anion	Compound or anion	Cation	H_2O	e =	Cation	Anion	Complex or compound	Compound or element
-0.39	Ni		$2Af^-$			-2e			$Ni(Af)_2$	
-0.39	Ni		$2Alan^-$			-2e			$Ni(Alan)_2$	
-0.39	Ni		$2Gl^-$			-2e			$Ni(Gl)_2$	
-0.39	Ni		$2Nta^{-3}$			-2e			$Ni(Nta)_2^{-4}$	
-0.39	Ni		OH^-			-2e			$NiOH^+$	
-0.38	Ni		$2Dge^-$			-2e			$Ni(Dge)_2$	
-0.38	Ni		$3En$			-2e			$Ni(En)_3^{+2}$	
-0.38	Ni		$Glgl^-$			-2e			$Ni(Glgl)^+$	
-0.38	Ni		$3Pn$			-2e			$Ni(Pn)_3^{+2}$	
-0.38	Ni		$2Tmen$			-2e			$Ni(Tmen)_2^{+2}$	
-0.37	Ni		$3Ist(b)^-$			-2e			$Ni(Ist(b))_3^-$	
-0.37	Ni		Mal^{-2}			-2e			$Ni(Mal)$	
-0.37	Ni		$3Met(b)^-$			-2e			$Ni(Met(b))_3^-$	
-0.37	Ni		$2Sald^-$			-2e			$Ni(Sald)_2$	
-0.37	Ni		$3Trop^-$			-2e			$Ni(Trop)_3^-$	
-0.36	Ni		$2Imdp^{-2}$			-2e			$Ni(Imdp)_2^{-2}$	
-0.36	Ni		$3Ist(a)^-$			-2e			$Ni(Ist(a))_3^-$	
-0.36	Ni		$3Met(a)^-$			-2e			$Ni(Met(a))_3^-$	
-0.36	Ni		$SSald^{-2}$			-2e			$Ni(SSald)$	
-0.35	Ni		$2Amac^{-3}$			-2e			$Ni(Amac)_2^{-4}$	
-0.35	Ni		$2Glgl^-$			-2e			$Ni(Glgl)_2$	
-0.35	Ni		Im			-2e			$Ni(Im)^{+2}$	
-0.35	Ni		$P_3O_9^{-3}$			-2e			$NiP_3O_9^-$	
-0.33	Ni		$2Im$			-2e			$Ni(Im)_2^{+2}$	
-0.33	Ni		NH_3			-2e			$Ni(NH_3)^{+2}$	
-0.33	Ni		N_2H_4			-2e			$Ni(N_2H_4)^{+2}$	
-0.33	Ni		$2SSald^{-2}$			-2e			$Ni(SSald)_2^{-2}$	
-0.32	Ni		$3AcAc^-$			-2e			$Ni(AcAc)_3^-$	
-0.32	Ni		$2NH_3$			-2e			$Ni(NH_3)_2^{+2}$	
-0.32	Ni		$2N_2H_4$			-2e			$Ni(N_2H_4)_2^{+2}$	
-0.32	Ni		SO_4^{-2}			-2e			$NiSO_4$	

EMF	Elem	Anion	Compound or anion	Cation	H_2O	e =	Cation	Anion	Complex or compound	Compound or element
-0.31	Ni		3Im			-2e			$Ni(Im)_3^{+2}$	
-0.31	Ni		$3N_2H_4$			-2e			$Ni(N_2H_4)_3^{+2}$	
-0.31	Ni		$S_2O_3^{-2}$			-2e			NiS_2O_3	
-0.30	Ni		Ac^-			-2e			$Ni(Ac)^+$	
-0.30	Ni		4Im			-2e			$Ni(Im)_4^{+2}$	
-0.30	Ni		$3NH_3$			-2e			$Ni(NH_3)_3^{+2}$	
-0.30	Ni		$4N_2H_4$			-2e			$Ni(N_2H_4)_4^{+2}$	
-0.30	Ni		$5N_2H_4$			-2e			$Ni(N_2H_4)_5^{+2}$	
-0.29	Ni		$4NH_3$			-2e			$Ni(NH_3)_4^{+2}$	
-0.29	Ni		$6N_2H_4$			-2e			$Ni(N_2H_4)_6^{+2}$	
-0.29	Ni		$2P_2O_7^{-4}$			-2e			$Ni(P_2O_7)_2^{-6}$	
-0.29	Ni		3Tmen			-2e			$Ni(Tmen)_3^{+2}$	
-0.28	Ni		CNS^-			-2e			$NiCNS^+$	
-0.28	Ni		5Im			-2e			$Ni(Im)_5^{+2}$	
-0.28	Ni		$2P_4O_{12}^{-4}$			-2e			$Ni(P_4O_{12})_2^{-6}$	
-0.27	Ni		Ac^-			-2e			$Ni(Ac)^+$	
-0.27	Ni		$2Ac^-$			-2e			$Ni(Ac)_2$	
-0.27	Ni		6Im			-2e			$Ni(Im)_6^{+2}$	
-0.27	Ni		$5NH_3$			-2e			$Ni(NH_3)_5^{+2}$	
-0.26	Ni		$2CNS^-$			-2e			$Ni(CNS)_2$	
-0.26	Ni		$3CNS^-$			-2e			$Ni(CNS)_3^-$	
-0.25	Ni					-2e	Ni^{+2}			
-0.25	Ni		Nac^-			-2e			$Ni(Nac)^+$	
-0.25	Ni		$6NH_3$			-2e			$Ni(NH_3)_6^{+2}$	
+0.49		$2OH^-$	$Ni(OH)_2$			-2e			NiO_2	$2H_2O$
+1.68	Ni				$2H_2O$	-2e	$4H^+$		NiO_2	

EMF	Elem	Anion	Compound or anion	Cation	H$_2$O	e =	Cation	Anion	Complex or compound	Compound or element
-2.50	No					-3e	No^{+3}			

EMF	Elem	Anion	Compound or anion	Cation	H_2O	e =	Cation	Anion	Complex or compound	Compound or element
-1.86	Np					-3e	Np^{+3}			
-1.75	Np		SO_4^{-2}			-4e			$NpSO_4^{+2}$	
-1.73	Np	$\cdot$	$2SO_4^{-2}$			-4e			$Np(SO_4)_2$	
-1.42	Np				$3H_2O$	-3e	$3H^+$		$Np(OH)_3$	
+0.15				Np^{+3}		-e	Np^{+4}			
+0.75				Np^{+4}	$2H_2O$	-e	NpO_2^+			$4H^+$
+1.15				NpO_2^+		-e	NpO_2^{+2}			

EMF	Elem	Anion	Compound or anion	Cation	H_2O	e =	Cation	Anion	Complex or compound	Compound or element
-2.93	H(g)		OH^-			-e			H_2O	
-0.93	H		OH^-			-e			H_2O	
-0.83	H_2		$2OH^-$			-2e			$2H_2O$	
-0.56		O_2^-				-e				O_2
-0.26		$2OH^-$	OH (g)			-e		HO_2^-	H_2O	
-0.25		$2OH^-$	OH (aq)			-e		HO_2^-	H_2O	
-0.13			HO_2			-e	H^+			O_2
-0.08		HO_2^-	OH^-			-2e			H_2O	O_2
+0.40			$4OH^-$			-4e			$2H_2O$	O_2
+0.41		OH^-	HO_2^-			-e		O_2^-	H_2O	
+0.68			H_2O_2(aq)			-2e	$2H^+$			O_2 (g)
+0.71			OH (g)		H_2O	-e	H^+		H_2O_2(aq)	
+0.88			$3OH^-$			-2e		HO_2^-	H_2O	
+1.19			$2H_2O$(g)			-4e	$4H^+$			O_2
+1.23			$2H_2O$ (1)			-4e	$4H^+$			O_2
+1.24	O_2		$2OH^-$			-2e			H_2O	O_3 (g)
+1.50			H_2O_2(aq)			-e	H^+		HO_2 (aq)	
+1.78					$2H_2O$	-2e	$2H^+$		H_2O_2	
+2.02			OH^-			-e			OH (g)	
+2.07	O_2				H_2O	-2e	$2H^+$			O_3
+2.42					H_2O	-2e	$2H^+$			O (g)
+2.85					H_2O	-e	H^+		OH	

EMF	Elem	Anion	Compound or anion	Cation	H_2O	e =	Cation	Anion	Complex or compound	Compound or element
-0.12	Os		$4OH^-$			-4e			$Os(OH)_4$	
+0.02	Os		$9OH^-$			-8e			$HOsO_5^-$	$4H_2O$
+0.69	Os				$2H_2O$	-2e	$4OH^+$		OsO_2	
+0.85	Os				$4H_2O$	-8e	$8H^+$		OsO_4	
+0.85	Os				$5H_2O$	-8e	$8H^+$		H_2OsO_5	

EMF	Elem	Anion	Compound or anion	Cation	H_2O	e =	Cation	Anion	Complex or compound	Compound or element
-2.05	P		$2OH^-$			-e		$H_2PO_2^-$		
-1.57		$H_2PO_2^-$	$3OH^-$			-2e		HPO_3^{-2}	$2H_2O$	
-1.12		HPO_3^{-2}	$3OH^-$			-2e		PO_4^{-3}	$2H_2O$	
-0.89		$3OH^-$	PH_3			-3e			$3H_2O$	P
-0.51	P				$2H_2O$	-e	H^+		H_3PO_2	
-0.50			$H_3PO_2(aq)$		H_2O	-2e	$2H^+$		$H_3PO_3(aq)$	
-0.45	P				$3H_2O$	-3e	$3H^+$		H_3PO_3	
-0.38	P				$4H_2O$	-5e	$5H^+$		H_3PO_4	
-0.37	P				$2H_2O$	-e	H^+		H_3PO_2	
-0.28			$H_3PO_3(aq)$		H_2O	-2e	$2H^+$		$H_3PO_4(aq)$	
-0.23	3H	PO_4^{-3}				-3e			H_3PO_4	
-0.21	2H	HPO_4^{-2}				-2e			H_3PO_4	
-0.18	2H	HPO_3^{-2}				-2e			H_3PO_3	
-0.14	3H	$HP_2O_7^{-3}$				-3e			$H_4P_2O_7$	
-0.13	4H	$P_2O_7^{-4}$				-4e			$H_4P_2O_7$	
-0.12	H	$H_2PO_4^-$				-e			H_3PO_4	
-0.11	H	$H_2PO_3^-$				-e			H_3PO_3	
-0.07	H	$H_2PO_2^-$				-e			H_3PO_2	
-0.06	H	$H_3P_2O_7^-$				-e			$H_4P_2O_7$	
-0.06	2H	$H_2P_2O_7^{-2}$				-2e			$H_4P_2O_7$	
-0.06			$PH_3\ (g)$			-3e	$3H^+$			P

EMF	Elem	Anion	Compound or anion	Cation	H_2O	e =	Cation	Anion	Complex or compound	Compound or element
-1.95	Pa					-3e	Pa^{+3}			
-1.70	Pa					-4e	Pa^{+4}			
-1.00				Pa^{+3}		-e	Pa^{+4}			
-1.00	Pa				$2H_2O$	-5e	PaO_2^+			$4H^+$
-0.50				Pa^{+3}	$2H_2O$	-e	$4H^+$		PaO_2	

EMF	Elem	Anion	Compound or anion	Cation	H_2O	e =	Cation	Anion	Complex or compound	Compound or element
-1.51			PbH_2			-2e	$2H^+$			Pb
-0.93	Pb		S^{-2}			-2e			PbS	
-0.71	Pb		$Data^{-4}$			-2e			$Pb(Data)^{-2}$	
-0.67	Pb		$Edta^{-4}$			-2e			$Pb(Edta)^{-2}$	
-0.58	Pb		$2OH^-$			-2e			PbO (r)	H_2O
-0.54	Pb		$3OH^-$			-2e		$HPbO_2^-$	H_2O	
-0.51	Pb		CO_3^{-2}			-2e			$PbCO_3$	
-0.48	Pb		Nta^{-3}			-2e			$Pb(Nta)^-$	
-0.44	Pb		Oxd^-			-2e			$Pb(Oxd)^+$	
-0.44	Pb		$Oxin^-$			-2e			$Pb(Oxin)^+$	
-0.41	Pb		$Himda^{-2}$			-2e			$Pb(Himda)$	
-0.41	Pb		$Ist(a)^-$			-2e			$Pb(Ist(a))^+$	
-0.41	Pb		$Met(a)^-$			-2e			$Pb(Met(a))^+$	
-0.41	Pb		$Met(b)^-$			-2e			$Pb(Met(b))^+$	
-0.40	Pb		$Ist(b)^-$			-2e			$Pb(Ist(b))^+$	
-0.38	Pb		$2N_3^-$			-2e			$Pb(N_3)_2$	
-0.37	Pb		$2I^-$			-2e			PbI_2	
-0.37	Pb		$2Oxd^-$			-2e			$Pb(Oxd)_2$	
-0.37	Pb		$2Oxin^-$			-2e			$Pb(Oxin)_2$	
-0.37	Pb		$Trop^-$			-2e			$Pb(Trop)^+$	
-0.36	Pb		SO_4^{-2}			-2e			$PbSO_4$	
-0.35	Pb		$2Ist(a)^-$			-2e			$Pb(Ist(a))_2$	
-0.35		SO_4^{-2}	Pb·Hg			-2e			$PbSO_4$	Hg
-0.34	Pb		Af^-			-2e			$Pb(Af)^+$	
-0.34	Pb		$2F^-$			-2e			PbF_2	
-0.33	Pb		$2Ist(b)^-$			-2e			$Pb(Ist(b))_2$	
-0.33	Pb		$2Met(a)^-$			-2e			$Pb(Met(a))_2$	
-0.33	Pb		$2Met(b)^-$			-2e			$Pb(Met(b))_2$	
-0.31	Pb		OH^-			-2e			$PbOH^+$	
-0.31	Pb		$2Trop^-$			-2e			$Pb(Trop)_2$	
-0.30	Pb		Cit^{-3}			-2e			$Pb(Cit)^-$	
-0.29	Pb		$2Af^-$			-2e			$Pb(Af)_2$	

EMF	Elem	Anion	Compound or anion	Cation	H_2O	e =	Cation	Anion	Complex or compound	Compound or element
-0.29	Pb		Gl^-			-2e			$Pb(Gl)^+$	
-0.28	Pb		$Alan^-$			-2e			$Pb(Alan)^+$	
-0.28	Pb		$2Br^-$			-2e			$PbBr_2$	
-0.28	Pb		$2S_2O_3^{-2}$			-2e			$Pb(S_2O_3)_2^{-2}$	
-0.27	Pb		$2Cl^-$			-2e			$PbCl_2$	
-0.25	Pb		$2Himda^{-2}$			-2e			$Pb(Himda)_2^{-2}$	
-0.23	Pb		$2Alan^-$			-2e			$Pb(Alan)_2$	
-0.23	Pb		$2Gl^-$			-2e			$Pb(Gl)_2$	
-0.23	Pb		$Glgl^-$			-2e			$Pb(Glgl)^+$	
-0.22	Pb		Dyp			-2e			$Pb(Dyp)^{+2}$	
-0.21	Pb		$2Glgl^-$			-2e			$Pb(Glgl)_2$	
-0.19	Pb		Ac^-			-2e			$Pb(Ac)^+$	
-0.18	Pb		Cl^-			-2e			$PbCl^+$	
-0.17	Pb		$3S_2O_3^{-2}$			-2e			$Pb(S_2O_3)_3^{-4}$	
-0.16	Pb		Br^-			-2e			$PbBr^+$	
-0.14	Pb		$2Ac^-$			-2e			$Pb(Ac)_2$	
-0.14	Pb		NO_3^-			-2e			$PbNO_3^+$	
-0.13	Pb					-2e	Pb^{+2}			
-0.13	Pb		$3Ac^-$			-2e			$Pb(Ac)_3^-$	
-0.13	Pb		Nac^-			-2e			$Pb(Nac)^+$	
-0.12	Pb		$4Ac^-$			-2e			$Pb(Ac)_4^{-2}$	
+0.25	Pb	$2OH^-$	PbO (r)			-2e			PbO_2	H_2O
+1.46				Pb^{+2}	$2H_2O$	-2e	$4H^+$		PbO_2	
+1.67				Pb^{+2}		-2e	Pb^{+4}			
+1.68			$PbSO_4$		$2H_2O$	-2e	$4H^+$	SO_4^{-2}	PbO_2	
+1.97	3Pb		$2PO_4^{-3}$			-6e			$Pb_3(PO_4)_2$	

EMF	Elem	Anion	Compound or anion	Cation	H_2O	e =	Cation	Anion	Complex or compound	Compound or element
-0.40	Pd		$4CN^-$			-2e			$Pd(CN)_4^{-2}$	
0.00	Pd		$4NH_3$			-2e			$Pd(NH_3)_4^{+2}$	
+0.05			Pd_2H			-e	H^+			2Pd
+0.07	Pd		$2OH^-$			-2e			$Pd(OH)_2$	
+0.14	Pd		$4SCN^-$			-2e			$Pd(SCN)_4^{-2}$	
+0.18	Pd		$4I^-$			-2e			PdI_4^{-2}	
+0.55	Pd		$5Cl^-$			-2e			$PdCl_5^{-3}$	
+0.57	Pd		$3Cl^-$			-2e			$PdCl_3^-$	
+0.60	Pd		$4Br^-$			-2e			$PdBr_4^{-2}$	
+0.62	Pd		$4Cl^-$			-2e			$PdCl_4^{-2}$	
+0.64	Pd		$2Cl^-$			-2e			$PdCl_2$	
+0.77	Pd		Cl^-			-2e			$PdCl^+$	
+0.99	Pd					-2e	Pd^{+2}			
+1.29		$2Cl^-$	$PdCl_4^{-2}$			-2e			$PdCl_6^{-4}$	

EMF	Elem	Anion	Compound or anion	Cation	H_2O	e =	Cation	Anion	Complex or compound	Compound or element
-2.84	Pm		$3OH^-$			-3e			$Pm(OH)_3$	
-2.42	Pm					-3e	Pm^{+3}			
-2.01	2Pm				$3H_2O$	-6e	$6H^+$		Pm_2O_3	

EMF	Elem	Anion	Compound or anion	Cation	H_2O	e =	Cation	Anion	Complex or compound	Compound or element
-1.00			H_2Po			-2e	$2H^+$			Po
-0.49	Po		$6OH^-$			-4e		PoO_3^{-2}	$3H_2O$	
+0.38	Po		$4Cl^-$			-2e		$PoCl_4^{-2}$		
+0.56	Po					-3e	Po^{+3}			
+0.65	Po					-2e	Po^{+2}			
+0.76	Po					-4e	Po^{+4}			
+0.80				Po^{+2}	$2H_2O$	-2e	$4H^+$		PoO_2	
+1.52			PoO_2		H_2O	-2e	$2H^+$		PoO_3	

EMF	Elem	Anion	Compound or anion	Cation	H_2O	e =	Cation	Anion	Complex or compound	Compound or element
-2.85	Pr		$3OH^-$			-3e			$Pr(OH)_3$	
-2.80	Pr		$Data^{-4}$			-3e			$Pr(Data)^-$	
-2.77	Pr		$Edta^{-4}$			-3e			$Pr(Edta)^-$	
-2.53	Pr		SO_4^{-2}			-3e			$PrSO_4^+$	
-2.46	Pr					-3e	Pr^{+3}			
-1.83	2Pr				$3H_2O$	-6e	$6H^+$		Pr_2O_3	
+2.76				Pr^{+3}	$2H_2O$	-e	$4H^+$		PrO_2	
+2.86				Pr^{+3}		-e	Pr^{+4}			

EMF	Elem	Anion	Compound or anion	Cation	H_2O	e =	Cation	Anion	Complex or compound	Compound or element
-0.33	Pt		H_2S (aq)			-2e	$2H^+$		PtS	
-0.33	Pt		H_2S (g)			-2e	$2H^+$		PtS	
+0.09	Pt		$4CN^-$			-2e			$Pt(CN)_4^{-2}$	
+0.10		$4OH^-$	$Pt(OH)_2$			-2e			$Pt(OH)_6^{-2}$	
+0.15	Pt		$2OH^-$			-2e			$Pt(OH)_2$	
+0.25	Pt		$4NH_3$			-2e			$Pt(NH_3)_4^{+2}$	
+0.40	Pt		$4I^-$			-2e			PtI_4^{-2}	
+0.58	Pt		$4Br^-$			-2e			$PtBr_4^{-2}$	
+0.61	Pt	$3Cl^-$	NH_3			-2e			$Pt(NH_3)Cl_3^-$	
+0.66	Pt	$5Cl^-$	NH_3			-4e			$Pt(NH_3)Cl_5^-$	
+0.68		$2Cl^-$	$PtCl_4^{-2}$			-2e			$PtCl_6^{-2}$	
+0.73	Pt		$4Cl^-$			-2e			$PtCl_{4.}^{-2}$	
+0.98	Pt				$2H_2O$	-2e	$2H^+$		$Pt(OH)_2$	
+0.98	Pt				H_2O	-2e	$2H^+$		PtO	
+1.10			$Pt(OH)_2$			-2e	$2H^+$		PtO_2	
+1.20	Pt					-2e	Pt^{+2}			

EMF	Elem	Anion	Compound or anion	Cation	H_2O	e =	Cation	Anion	Complex or compound	Compound or element
-2.42	Pu		$3OH^-$			-3e			$Pu(OH)_3$	
-2.03	Pu					-3e	Pu^{+3}			
-1.19	Pu		OH^-			-4e			$PuOH^{+3}$	
-1.18	Pu		Tf^-			-4e			$Pu(Tf)^{+3}$	
-1.16	Pu		F^-			-4e			PuF^{+3}	
-1.11	Pu		SO_4^{-2}			-4e			$PuSO_4^{+2}$	
-1.07	Pu		NO_3^-			-4e			$PuNO_3^{+3}$	
-1.04	Pu		Cl^-			-4e			$PuCl^{+3}$	
-0.96		OH^-	$Pu(OH)_3$			-e			$Pu(OH)_4$	
+0.23		OH^-	PuO_2OH			-e			$PuO_2(OH)_2$	
+0.93				PuO_2^+		-e	PuO_2^{+2}			
+0.97				Pu^{+3}		-e	Pu^{+4}			
+1.04				Pu^{+4}	$2H_2O$	-2e	PuO_2^{+2}		$4H^+$	
+1.15				Pu^{+4}	$2H_2O$	-e	PuO_2^+		$4H^+$	

EMF	Elem	Anion	Compound or anion	Cation	H_2O	e =	Cation	Anion	Complex or compound	Compound or element
-4.00	Ra					-e	Ra^+			
-2.98	Ra		Cit^{-3}			-2e			$Ra(Cit)^-$	
-2.95	Ra		$Aspa^{-2}$			-2e			$Ra(Aspa)$	
-2.95	Ra		Suc^{-2}			-2e			$Ra(Suc)$	
-2.92	Ra					-2e	Ra^{+2}			
-1.32	Ra				H_2O	-2e	$2H^+$			RaO

EMF	Elem	Anion	Compound or anion	Cation	H_2O	e =	Cation	Anion	Complex or compound	Compound or element
-2.93	Rb					-e	Rb^+			
-1.30			RbH			-2e	Rb^+			H^+
+0.32			RbH			-e	H^+			Rb

EMF	Elem	Anion	Compound or anion	Cation	H_2O	e =	Cation	Anion	Complex or compound	Compound or element
-0.59		$4OH^-$	ReO_2			-3e		ReO_4^-	$2H_2O$	
-0.58	Re	$4OH^-$				-4e			ReO_2	$2H_2O$
-0.58	Re		$8OH^-$			-7e		ReO_4^-		$4H_2O$
-0.40		Re^-				-e				Re
+0.13		Re^-				-4e	Re^{+3}			
+0.23	2Re				$3H_2O$	-6e	$6H^+$		Re_2O_3	
+0.25	Re				$2H_2O$	-4e	$4H^+$		ReO_2	
+0.30	Re					-3e	Re^{+3}			
+0.36	Re				$4H_2O$	-7e	$8H^+$	ReO_4^-		
+0.51			ReO_2		$2H_2O$	-3e	$4H^+$	ReO_4^-		

EMF	Elem	Anion	Compound or anion	Cation	H_2O	e =	Cation	Anion	Complex or compound	Compound or element
0.00	Rh		$3OH^-$			-3e			$Rh(OH)_3$	
+0.04	2Rh		$6OH^-$			-6e			Rh_2O_3	$3H_2O$
-0.43	Rh		$6Cl^-$			-3e		$RhCl_6^{-3}$		
+0.60	Rh					-e	Rh^+			
+0.60	Rh					-2e	Rh^{+2}			
+0.60				Rh^+		-e	Rh^{+2}			
+0.80	2Rh				H_2O	-2e	$2H^+$		Rh_2O	
+0.80	Rh					-3e	Rh^{+3}			
+0.81	Rh				H_2O	-2e	$2H^+$		RhO	
+0.87	2Rh				$3H_2O$	-6e	$6H^+$		Rh_2O_3	
+1.10	Rh				$4H_2O$	-6e	$8H^+$	RhO_4^{-2}		
+1.20				Rh^{+2}		-e	Rh^{+3}			
+1.20			$(RhCl_6)^{-3}$			-e			$(RhCl_6)^{-2}$	

EMF	Elem	Anion	Compound or anion	Cation	H_2O	e =	Cation	Anion	Complex or compound	Compound or element
-0.15	Ru		$4OH^-$			-4e			$Ru(OH)_4$	
+0.21			$(RuNh)^{+2}$			-e			$(RuNh)^{+3}$	
+0.25				Ru^{+2}		-e	Ru^{+3}			
+0.46	Ru					-2e	Ru^{+2}			
+0.60	Ru		$5Cl^-$			-3e		$RuCl_3^{-2}$		
+0.60			RuO_4^{-2}			-e		RuO_4^-		
+0.68	Ru				$4H_2O$	-4e	$4H^+$		$Ru(OH)_4$	
+0.74	2Ru				$3H_2O$	-6e	$6H^+$		Ru_2O_3	
+0.86			$(RuCy)^{-4}$			-e			$(RuCy)^{-3}$	

.

EMF	Elem	Anion	Compound or anion	Cation	H_2O	e =	Cation	Anion	Complex or compound	Compound or element
-1.12		$S_2O_4^{-2}$	$4OH^-$			-2e		$2SO_3^{-2}$	$2H_2O$	
-0.93		SO_3^{-2}	$2OH^-$			-2e		SO_4^{-2}	H_2O	
-0.57		$S_2O_3^{-2}$	$6OH^-$			-4e		$2SO_3^{-2}$	$3H_2O$	
-0.45			S^{-2}			-2e				S
-0.41	H		HS^-			-e			H_2S	
-0.41	2H		S^{-2}			-2e			H_2S	
-0.22			$S_2O_6^{-2}$		$2H_2O$	-2e	$4H^+$	$2SO_4^{-2}$		
-0.21	2H		SO_3^{-2}			-2e			H_2SO_3	
-0.11	H		HSO_3^-			-e			H_2SO_3	
-0.08			$H_2SO_4^-$		$2H_2O$	-2e	H^+		$2H_2SO_3$	
-0.07	H		SO_4^{-2}			-e			HSO_4^-	
-0.06	2H		SO_4^{-2}			-2e			H_2SO_4	
+0.08			$2S_2O_3^{-2}$			-2e		$S_4O_6^{-2}$		
+0.14			H_2S (aq)			-2e	$2H^+$			S
+0.17			H_2SO_3		H_2O	-2e	$4H^+$	SO_4^{-2}		
+0.18	H		HSO_4^-			-e			H_2SO_4	
+0.36	S				$4H_2O$	-6e	$8H^+$	SO_4^{-2}		
+0.40			$S_2O_3^{-2}$		$3H_2O$	-4e	$2H^!$		$2H_2SO_3$	
+0.45	S				$3H_2O$	-4e	$4H^+$		H_2SO_3	
+0.47	2S				$3H_2O$	-4e	$6H^+$	$S_2O_3^{-2}$		
+0.51			$S_4O_6^{-2}$		$6H_2O$	-6e	$4H^+$		$4H_2SO_3$	
+0.57			$2H_2SO_3$			-2e	$4H^+$	$S_2O_6^{-2}$		
+1.23	2S		$2Cl^-$			-2e			S_2Cl_2	
+2.01			$2SO_4^{-2}$			-2e		$S_2O_8^{-2}$		

EMF	Elem	Anion	Compound or anion	Cation	H_2O	e =	Cation	Anion	Complex or compound	Compound or element
-0.66	Sb		$4OH^-$			-3e		SbO_2^-	$2H_2O$	
-0.51			SbH_3 (g)			-3e	$3H^+$			Sb
+0.15	2Sb				$3H_2O$	-6e	$6H^+$		Sb_2O_3	
+0.17	Sb		$4Cl^-$			-3e		$SbCl_4^-$		
+0.48			Sb_2O_4 (c)		H_2O	-2e	$2H^+$		Sb_2O_5 (c)	
+0.58				$2SbO^+$	$3H_2O$	-4e	$6H^+$		Sb_2O_5 (c)	

EMF	Elem	Anion	Compound or anion	Cation	H_2O	e =	Cation	Anion	Complex or compound	Compound or element
-2.61	Sc		$3OH^-$			-3e			$Sc(OH)_3$	
-2.43	Sc		$3F^-$			-3e			ScF_3	
-2.33	Sc		$2F^-$			-3e			ScF_2^+	
-2.32	Sc		OH^-			-3e			$ScOH^{+2}$	
-2.24	Sc		$AcAc^-$			-3e			$Sc(AcAc)^{+2}$	
-2.24	Sc		F^-			-3e			ScF^{+2}	
-2.22	Sc		$2AcAc^-$			-3e			$Sc(AcAc)_2^+$	
-2.08	Sc					-3e	Sc^{+3}			
-1.59	2Sc				$3H_2O$	-6e	$6H^+$		Sc_2O_3	

EMF	Elem	Anion	Compound or anion	Cation	H_2O	e =	Cation	Anion	Complex or compound	Compound or element
-0.92		Se^{-2}				-2e				Se
-0.40			$H_2Se(aq)$			-2e	$2H^+$			Se
-0.37	Se		$6OH^-$			-4e		SeO_3^{-2}	$3H_2O$	
-0.24	H		HSe^-			-e			H_2Se	
-0.06	2H		SeO_4^{-2}			-2e			H_2SeO_4	
+0.05		$2OH^-$	SeO_3^{-2}			-2e		SeO_4^{-2}		H_2O
+0.18	H		$HSeO_4^-$			-e			H_2SeO_4	
+0.74	Se				$3H_2O$	-4e	$4H^+$		$H_2SeO_3(aq)$	
+1.15			H_2SeO_3		H_2O	-2e	$4H^+$		SeO_4^{-2}	

EMF	Elem	Anion	Compound or anion	Cation	H_2O	e =	Cation	Anion	Complex or compound	Compound or element
-1.70	Si		$6OH^-$			-4e		SiO_3^{-2}	$3H_2O$	
-1.24	Si		$6F^-$			-4e			SiF_6^{-2}	
-0.86	Si				$2H_2O$	-4e	$4H^+$		SiO_2	
-0.78	Si				$3H_2O$	-4e	$4H^+$		H_2SiO_3	
+0.10			SiH_4 (g)			-4e	$4H^+$			Si

EMF	Elem	Anion	Compound or anion	Cation	H_2O	e =	Cation	Anion	Complex or compound	Compound or element
-3.12	Sm					-2e	Sm^{+2}			
-2.83	Sm		$3OH^-$			-3e			$Sm(OH)_3$	
-2.77	Sm		$Data^{-4}$			-3e			$Sm(Data)^-$	
-2.74	Sm		$Edta^{-4}$			-3e			$Sm(Edta)^-$	
-2.53	Sm		$AcAc^-$			-3e			$Sm(AcAc)^{+2}$	
-2.50	Sm		$2AcAc^-$			-3e			$Sm(AcAc)_2^+$	
-2.48	Sm		SO_4^{-2}			-3e			$SmSO_4^+$	
-2.47	Sm		$3AcAc^-$			-3e			$Sm(AcAc)_3$	
-2.41	Sm					-3e	Sm^{+3}			
-2.00	2Sm				$3H_2O$	-6e	$6H^+$		Sm_2O_3	
-1.15				Sm^{+2}		-e	Sm^{+3}			

EMF	Elem	Anion	Compound or anion	Cation	H_2O	e =	Cation	Anion	Complex or compound	Compound or element
-1.07			SnH_4			-4e	$4H^+$			Sn
-0.93		$HSnO_2^-$	$3OH^-$		H_2O	-2e			$Sn(OH)_6^{-2}$	
-0.91	Sn		$3OH^-$			-2e		$HSnO_2^-$	H_2O	
-0.87	Sn		S^{-2}			-2e			SnS	
-0.43	Sn		OH^-			-2e			$SnOH^+$	
-0.25	Sn		$6F^-$			-4e			SnF_6^{-2}	
-0.17	Sn		Cl^-			-2e			$SnCl^+$	
-0.16	Sn		Br^-			-2e			$SnBr^+$	
-0.16	Sn		$2Cl^-$			-2e			$SnCl_2$	
-0.15	Sn		$2Br^-$			-2e			$SnBr_2$	
-0.15	Sn		$3Br^-$			-2e			$SnBr_3^-$	
-0.14	Sn					-2e	Sn^{+2}			
-0.14	Sn		$3Cl^-$			-2e			$SnCl_3^-$	
-0.11	Sn				$2H_2O$	-4e	$4H^+$		SnO_2	
-0.10	Sn				H_2O	-2e	$2H^+$		SnO	
+0.15				Sn^{+2}		-2e	Sn^{+4}			

EMF	Elem	Anion	Compound or anion	Cation	H_2O	e =	Cation	Anion	Complex or compound	Compound or element
-4.10	Sr					-e	Sr^+			
-3.15	Sr		$Edta^{-4}$			-2e			$Sr(Edta)^{-2}$	
-3.12	Sr		$Amac^{-3}$			-2e			$Sr(Amac)^-$	
-3.09	Sr		Nta^{-3}			-2e			$Sr(Nta)^-$	
-3.04	Sr		$P_4O_{12}^{-4}$			-2e			$SrP_4O_{12}^{-2}$	
-3.04	Sr		$Tmta^{-4}$			-2e			$Sr(Tmta)^{-2}$	
-2.99	Sr		$P_3O_9^{-3}$			-2e			$SrP_3O_9^-$	
-2.99	Sr		Saa^{-3}			-2e			$Sr(Saa)^-$	
-2.98	Sr		Cit^{-3}			-2e			$Sr(Cit)^-$	
-2.97	Sr		Ox^{-2}			-2e			$Sr(Ox)$	
-2.97	Sr		$(FeCy)^{-3}$			-2e			$Sr(FeCy)^-$	
-2.95	Sr		$S_2O_3^{-2}$			-2e			SrS_2O_3	
-2.94	Sr		$Tart^{-2}$			-2e			$Sr(Tart)$	
-2.93	Sr		Ap^{-2}			-2e			$Sr(Ap)$	
-2.93	Sr		$Aspa^{-2}$			-2e			$Sr(Aspa)$	
-2.93	Sr		Mal^{-2}			-2e			$Sr(Mal)$	
-2.92	Sr		$Glac^-$			-2e			$Sr(Glac)^+$	
-2.92	Sr		Glu^-			-2e			$Sr(Glu)^+$	
-2.92	Sr		IO_3^-			-2e			$SrIO_3^+$	
-2.92	Sr		Suc^{-2}			-2e			$Sr(Suc)$	
-2.91	Sr		Gl^-			-2e			$Sr(Gl)^+$	
-2.91	Sr		Gly^-			-2e			$Sr(Gly)^+$	
-2.91	Sr		Lac^-			-2e			$Sr(Lac)^+$	
-2.91	Sr		NO_3^-			-2e			$SrNO_3^+$	
-2.91	Sr		OH^-			-2e			$SrOH^+$	
-2.90	Sr		Ac^-			-2e			$Sr(Ac)^+$	
-2.90	Sr		Bu^-			-2e			$Sr(Bu)^+$	
-2.90	Sr		Pr^-			-2e			$Sr(Pr)^+$	
-2.89	Sr					-2e	Sr^{+2}			
-2.88	Sr		$2OH^-$			-2e			$Sr(OH)_2$	

EMF	Elem	Anion	Compound or anion	Cation	H_2O	e =	Cation	Anion	Complex or compound	Compound or element
-1.97	Sr				H_2O	-2e	$2H^+$		SrO	
-1.09			SrH_2			-4e	Sr^{+2}			$2H^+$
+0.72			SrH_2			-2e	$2H^+$			Sr

EMF	Elem	Anion	Compound or anion	Cation	H_2O	e =	Cation	Anion	Complex or compound	Compound or element
-0.81	2Ta				$5H_2O$	-10e	$10H^+$			Ta_2O_5

EMF	Elem	Anion	Compound or anion	Cation	H_2O	e =	Cation	Anion	Complex or compound	Compound or element
-2.79	Tb		$3OH^-$			-3e			$Tb(OH)_3$	
-2.77	Tb		$Data^{-4}$			-3e			$Tb(Data)^-$	
-2.73	Tb		$Edta^{-4}$			-3e			$Tb(Edta)^-$	
-2.39	Tb					-3e	Tb^{+3}			
-2.00	2Tb				$3H_2O$	-6e	$6H^+$			Tb_2O_3

EMF	Elem	Anion	Compound or anion	Cation	H_2O	e =	Cation	Anion	Complex or compound	Compound or element
+0.40	Tc					-2e	Tc^{+2}			
+0.60				Tc^{+2}	$2H_2O$	-2e	$4H^+$		TcO_2	
+0.70			TcO_2		$2H_2O$	-3e	$4H^+$	TcO_4^-		

EMF	Elem	Anion	Compound or anion	Cation	H_2O	e =	Cation	Anion	Complex or compound	Compound or element
-1.14		Te^{-2}				-2e				Te
-0.74			$H_2Te(aq)$			-2e	$2H^+$			Te
-0.72			H_2Te (g)			-2e	$2H^+$			Te
-0.57	Te		$6OH^-$			-4e		TeO_3^{-2}	$3H_2O$	
-0.45	H		$HTeO_4^-$			-e			H_2TeO_4	
-0.18	H		HTe^-			-e			H_2Te	
+0.40		$2OH^-$	TeO_3^{-2}			-2e		TeO_4^{-2}		H_2O
+0.53	Te				$2H_2O$	-4e	$4H^+$		TeO_2 (c)	
+0.56	Te				$2H_2O$	-4e	$3H^+$		$TeOOH^+$	
+0.57	Te					-4e	Te^{+4}			
+1.02			TeO_2		$4H_2O$	-2e	$2H^+$		H_6TeO_6(c)	

EMF	Elem	Anion	Compound or anion	Cation	H_2O	e =	Cation	Anion	Complex or compound	Compound or element
-2.48	Th		$4OH^-$			-4e			$Th(OH)_4$	
-2.03	Th		$AcAc^-$			-4e			$Th(AcAc)^{+3}$	
-2.01	Th		$2AcAc^-$			-4e			$Th(AcAc)_2^{+2}$	
-2.01	Th		F^-			-4e			ThF^{+3}	
-2.01	Th		$3IO_3^-$			-4e			$Th(IO_3)_3^+$	
-2.01	Th		Tf^-			-4e			$Th(Tf)^{+3}$	
-1.99	Th		$3AcAc^-$			-4e			$Th(AcAc)_3^+$	
-1.99	Th		$2F^-$			-4e			ThF_2^{+2}	
-1.97	Th		$3F^-$			-4e			ThF_3^+	
-1.97	Th		$2IO_3^-$			-4e			$Th(IO_3)_2^{+2}$	
-1.96	Th		$4AcAc^-$			-4e			$Th(AcAc)_4$	
-1.96	Th		$H_2PO_4^-$			-4e			$ThH_2PO_4^{+3}$	
-1.96	Th		$2H_2PO_4^-$			-4e			$Th(H_2PO_4)_2^{+2}$	
-1.95	Th		SO_4^{-2}			-4e			$ThSO_4^{+2}$	
-1.94	Th		IO_3^-			-4e			$ThIO_3^{+3}$	
-1.93	Th		H_3PO_4			-4e			$Th(H_3PO_4)^{+4}$	
-1.93	Th		$2SO_4^{-2}$			-4e			$Th(SO_4)_2$	
-1.91	Th		BrO_3^-			-4e			$ThBrO_3^{+3}$	
-1.91	Th		NO_3^-			-4e			$ThNO_3^{+3}$	
-1.91	Th		OH^-			-4e			$ThOH^{+3}$	
-1.90	Th					-4e	Th^{+4}			
-1.90	Th		$2BrO_3^-$			-4e			$Th(BrO_3)_2^{+2}$	
-1.90	Th		Cl^-			-4e			$ThCl^{+3}$	
-1.90	Th		ClO_3^-			-4e			$ThClO_3^{+3}$	
-1.89	Th		$3Cl^-$			-4e			$ThCl_3^+$	
-1.89	Th		$4Cl^-$			-4e			$ThCl_4$	
-1.88	Th		$2Cl^-$			-4e			$ThCl_2^{+2}$	

EMF	Elem	Anion	Compound or anion	Cation	H_2O	e =	Cation	Anion	Complex or compound	Compound or element
-1.63	Ti					-2e	Ti^{+2}			
-1.19	Ti		$6F^-$			-4e			TiF_6^{-2}	
-0.88	Ti				H_2O	-4e	TiO_2^+		$2H^+$	
-0.79			$Ti(OH)_3$			-e	H^+		TiO_2	H_2O
-0.37				Ti^{+2}		-e	Ti^{+3}			
0.00				Ti^{+3}		-e	Ti^{+4}			
+0.10				Ti^{+3}	H_2O	-e	$2H^+$		TiO^{+2}	

EMF	Elem	Anion	Compound or anion	Cation	H₂O	e =	Cation	Anion	Complex or compound	Compound or element
-1.87			TlH			-e	H^+			Tl
-1.17	Tl		OH^-			-e			TlOH	
-1.06	2Tl		CrO_4^{-2}			-2e			Tl_2CrO_4	
-0.90	2Tl		S^{-2}			-2e			Tl_2S	
-0.75	Tl		I^-			-e			TlI	
-0.66	Tl		Br^-			-e			TlBr	
-0.55	Tl		N_3^-			-e			TlN_3	
-0.45	Tl		$S_2O_3^{-2}$			-e		$TlS_2O_3^-$		
-0.44	Tl		$P_2O_7^{-4}$			-e			$TlP_2O_7^{-3}$	
-0.39	Tl		CNS^-			-e			TlCNS	
-0.38	Tl		Cl^-			-e			TlCl	
-0.37	Tl		IO_3^-			-e			$TlIO_3$	
-0.35	Tl		F^-			-e			TlF	
-0.35	Tl		$2P_2O_7^{-4}$			-e			$Tl(P_2O_7)_2^{-7}$	
-0.34	Tl					-e	Tl^+			
-0.34	Tl		OH^-			-e			TlOH (c)	
-0.29	Tl		NH_3			-e			$TlNH_3^+$	
-0.05		$2OH^-$	TlOH			-2e			$Tl(OH)_3$	
+0.63	Tl		OH^-			-3e			$Tl(OH)^{+2}$	
+0.65	Tl		$2OH^-$			-3e			$Tl(OH)_2^+$	
+0.72	Tl		Br^-			-3e			$TlBr^{+2}$	
+0.75	Tl		Cl^-			-3e			$TlCl^{+2}$	
+0.77	Tl		$2Br^-$			-3e			$TlBr_2^+$	
+0.80	Tl		$2Cl^-$			-3e			$TlCl_2^+$	
+0.82	Tl		$3Br^-$			-3e			$TlBr_3$	
+0.86	Tl		$4Br^-$			-3e			$TlBr_4^-$	
+0.86	Tl		$4Cl^-$			-3e			$TlCl_4^-$	
+0.87	Tl		$3Cl^-$			-3e			$TlCl_3$	
+0.91	Tl		NO_3^-			-3e			$TlNO_3^{+2}$	
+1.25				Tl^+		-2e	Tl^{+3}			

EMF	Elem	Anion	Compound or anion	Cation	H_2O	e =	Cation	Anion	Complex or compound	Compound or element
-2.74	Tm		$3OH^-$			-3e			$Tm(OH)_3$	
-2.28	Tm					-3e	Tm^{+3}			
-1.91	2Tm				$3H_2O$	-6e	$6H^+$		Tm_2O_3	

EMF	Elem	Anion	Compound or anion	Cation	H_2O	e =	Cation	Anion	Complex or compound	Compound or element
-2.58	U		OH^-			-4e			UOH^{+3}	
-2.51	U		Tf^-			-4e			$U(Tf)^{+3}$	
-2.45	U		SO_4^{-2}			-4e			USO_4^{+2}	
-2.43	U		$2SO_4^{-2}$			-4e			$U(SO_4)_2$	
-2.42	U		CNS^-			-4e			$UCNS^{+3}$	
-2.41	U		Cl^-			-4e			UCl^{+3}	
-2.41	U		$2CNS^-$			-4e			$U(CNS)_2^{+2}$	
-2.40	U		Br^-			-4e			UBr^{+3}	
-2.39	U		$4OH^-$			-4e			$2H_2O$	UO_2
-2.20		OH^-	$U(OH)_3$			-e			$U(OH)_4$	
-2.17	U		$3OH^-$			-3e			$U(OH)_3$	
-1.79	U					-3e	U^{+3}			
-1.62		$4OH^-$	$U(OH)_4$	$2Na^+$		-2e			Na_2UO_4	$4H_2O$
-0.62			$mOxin^-$	UO_2^+		-e			$UO_2(mOxin)^+$	
-0.62			$meOxin^-$	UO_2^+		-e			$UO_2(meOxin)^+$	
-0.62			$Oxin^-$	UO_2^+		-e			$UO_2(Oxin)^+$	
-0.61				U^{+3}		-e	U^{+4}			
-0.59			$MeOxin^-$	UO_2^+		-e			$UO_2(MeOxin)^+$	
-0.55			Coy^-	UO_2^+		-e			$UO_2(Coy)^+$	
-0.53			$2meOxin^-$	UO_2^+		-e			$UO_2(meOxin)_2$	
-0.52			$2Oxin^-$	UO_2^+		-e			$UO_2(Oxin)_2$	
-0.51			$2mOxin^-$	UO_2^+		-e			$UO_2(mOxin)_2$	
-0.51			Oxd^-	UO_2^+		-e			$UO_2(Oxd)^+$	
-0.50			$2MeOxin^-$	UO_2^+		-e			$UO_2(MeOxin)_2$	
-0.48			Has^-	UO_2^+		-e			$UO_2(Has)^+$	
-0.48			$Mcin^-$	UO_2^+		-e			$UO_2(Mcin)^+$	
-0.47			$DmHas^-$	UO_2^+		-e			$UO_2(DmHas)^+$	
-0.46			Cin^-	UO_2^+		-e			$UO_2(Cin)^+$	
-0.45			Hox^-	UO_2^+		-e			$UO_2(Hox)^+$	
-0.45			$MpHas^-$	UO_2^+		-e			$UO_2(MpHas)^+$	
-0.42			$2Oxd^-$	UO_2^+		-e			$UO_2(Oxd)_2$	
-0.41			$2Has^-$	UO_2^+		-e			$UO_2(Has)_2$	

EMF	Elem	Anion	Compound or anion	Cation	H_2O	e =	Cation	Anion	Complex or compound	Compound or element
-0.41			2MpHas⁻	UO_2^+		-e			$UO_2(MpHas)_2$	
-0.40			AcAc⁻	UO_2^+		-e			$UO_2(AcAc)^+$	
-0.39			2Coy⁻	UO_2^+		-e			$UO_2(Coy)_2$	
-0.39			2Hox⁻	UO_2^+		-e			$UO_2(Hox)_2$	
-0.39			2Mcin⁻	UO_2^+		-e			$UO_2(Mcin)_2$	
-0.38			2DmHas⁻	UO_2^+		-e			$UO_2(DmHas)_2$	
-0.37			2Cin⁻	UO_2^+		-e			$UO_2(Cin)_2$	
-0.33			2AcAc⁻	UO_2^+		-e			$UO_2(AcAc)_2$	
-0.22			F⁻	UO_2^+		-e			UO_2F^+	
-0.15			2F⁻	UO_2^+		-e			UO_2F_2	
-0.10			3F⁻	UO_2^+		-e			$UO_2F_3^-$	
-0.09			Gly⁻	UO_2^+		-e			$UO_2(Gly)^+$	
-0.05			SO_4^{-2}	UO_2^+		-e			UO_2SO_4	
-0.04			2Gly⁻	UO_2^+		-e			$UO_2(Gly)_2$	
-0.03			4F⁻	UO_2^+		-e			$UO_2F_4^{-2}$	
-0.02			3Gly⁻	UO_2^+		-e			$UO_2(Gly)_3^-$	
0.00			$3SO_4^{-2}$	UO_2^+		-e			$UO_2(SO_4)_3^{-4}$	
+0.01			CNS⁻	UO_2^+		-e			UO_2CNS^+	
+0.01			$2SO_4^{-2}$	UO_2^+		-e			$UO_2(SO_4)_2^{-2}$	
+0.02			3CNS⁻	UO_2^+		-e			$UO_2(CNS)_3^-$	
+0.05				UO_2^+		-e	UO_2^{+2}			
+0.05			2CNS⁻	UO_2^+		-e			$UO_2(CNS)_2$	
+0.07			Br⁻	UO_2^+		-e			UO_2Br^+	
+0.07			Cl⁻	UO_2^+		-e			UO_2Cl^+	
+0.09			NO_3^-	UO_2^+		-e			$UO_2NO_3^+$	
+0.26			UH_3			-3e	$3H^+$			U
+0.33				U^{+4}	$2H_2O$	-2e	$4H^+$		UO_2^{+2}	
+0.62				U^{+4}	$2H_2O$	-e	$4H^+$		UO_2^+	

EMF	Elem	Anion	Compound or anion	Cation	H_2O	e =	Cation	Anion	Complex or compound	Compound or element
-1.67	V		OH^-			-3e	VOH^{+2}			
-1.66	V		$2OH^-$			-3e	$V(OH)_2^+$			
-1.49	V		CNS^-			-3e			$VCNS^{+2}$	
-1.19	V					-2e	V^{+2}			
-1.15	6V		$33OH^-$			-30e		$HV_6O_{17}^{-3}$	$16H_2O$	
-0.26				V^{+2}		-e	V^{+3}			
-0.25	V				$4H_2O$	-5e	$4H^+$		$V(OH)_4^+$	
+0.36				V^{+3}	H_2O	-e	$2H^+$		VO_2^+	
+1.00				VO^{+2}	$3H_2O$	-e	$V(OH)_4^+$		$2H^+$	

EMF	Elem	Anion	Compound or anion	Cation	H_2O	e =	Cation	Anion	Complex or compound	Compound or element
-1.05	W		$8OH^-$			-6e		WO_4^{-2}		$4H_2O$
-0.09	W				$3H_2O$	-6e	$6H^+$			WO_3
-0.12	W				$2H_2O$	-4e	$4H^+$			WO_2
+0.46			$(W(CN)_8)^{-4}$			-e				$(W(CN)_8)^{-3}$

EMF	Elem	Anion	Compound or anion	Cation	H_2O	e =	Cation	Anion	Complex or compound	Compound or element
+0.90		$4OH^-$	$HXeO_4^-$			-2e		$HXeO_6^{-3}$	$2H_2O$	
+0.90	Xe	$7OH^-$				-6e		$HXeO_4^-$	$3H_2O$	
+1.80	Xe				$3H_2O$	-6e	$6H^+$		XeO_3	
+3.00			XeO_3		$3H_2O$	-2e	$2H^+$		H_4XeO_6	

EMF	Elem	Anion	Compound or anion	Cation	H_2O	e =	Cation	Anion	Complex or compound	Compound or element
-2.81	Y		$3OH^-$			-3e			$Y(OH)_3$	
-2.75	Y		$Data^{-4}$			-3e			$Y(Data)^-$	
-2.71	Y		$Edta^{-4}$			-3e			$Y(Edta)^-$	
-2.50	Y		$AcAc^-$			-3e			$Y(AcAc)^{+2}$	
-2.48	Y		$Oxac^{-2}$			-3e			$Y(Oxac)^+$	
-2.46	Y		$2AcAc^-$			-3e			$Y(AcAc)_2^+$	
-2.45	Y		$2Oxac^{-2}$			-3e			$Y(Oxac)_2^-$	
-2.44	Y		SO_4^{-2}			-3e			YSO_4^+	
-2.43	Y		$3AcAc^-$			-3e			$Y(AcAc)_3$	
-2.37	Y					-3e	Y^{+3}			
-1.98	Y				$3H_2O$	-3e	$3H^+$		$Y(OH)_3$	
-1.68	2Y				$3H_2O$	-6e	$6H^+$		Y_2O_3	

EMF	Elem	Anion	Compound or anion	Cation	H_2O	e =	Cation	Anion	Complex or compound	Compound or element
-2.80	Yb					-2e	Yb^{+2}			
-2.73	Yb		$3OH^-$			-3e				$Yb(OH)_3$
-2.69	Yb		$Data^{-4}$			-3e				$Yb(Data)^-$
-2.64	Yb		$Edta^{-4}$			-3e				$Yb(Edta)^-$
-2.41	Yb		Ox^{-2}			-3e				$Yb(OH)^+$
-2.36	Yb		$2Ox^{-2}$			-3e				$Yb(Ox)_2^-$
-2.34	Yb		SO_4^{-2}			-3e				$YbSO_4^+$
-2.27	Yb					-3e	Yb^{+3}			
-1.90	2Yb				$3H_2O$	-6e	$6H^+$			Yb_2O_3
-1.21				Yb^{+2}		-e	Yb^{+3}			

EMF	Elem	Anion	Compound or anion	Cation	H_2O	e =	Cation	Anion	Complex or compound	Compound or element
-1.41	Zn		S^{-2}			-2e			ZnS	
-1.31	Zn		Data^{-4}			-2e			Zn(Data)$^{-2}$	
-1.26	Zn		4CN$^-$			-2e			Zn(CN)$_4^{-2}$	
-1.25	Zn		Edta^{-4}			-2e			Zn(Edta)$^{-2}$	
-1.25	Zn		2OH$^-$			-2e			Zn(OH)$_2$	
-1.22	Zn		4OH$^-$			-2e		ZnO_2^{-2}	$2H_2O$	
-1.19	Zn		Hed^{-3}			-2e			Zn(Hed)$^-$	
-1.19	Zn		Tate			-2e			Zn(Tate)$^{+2}$	
-1.12	Zn		Teta			-2e			Zn(Teta)$^{+2}$	
-1.07	Zn		Nta^{-3}			-2e			Zn(Nta)$^-$	
-1.06	Zn		CO_3^{-2}			-2e			ZnCO$_3$	
-1.06	Zn		Ndap^{-3}			-2e			Zn(Ndap)$^-$	
-1.05	Zn		Oxin$^-$			-2e			Zn(Oxin)$^+$	
-1.04	Zn		4NH$_3$(aq)			-2e			Zn(NH$_3$)$_4^{+2}$	
-1.04	Zn		meOxin$^-$			-2e			Zn(meOxin)$^+$	
-1.02	Zn		Deta			-2e			Zn(Deta)$^{+2}$	
-1.02	Zn		Ist(a)$^-$			-2e			Zn(Ist(a))$^+$	
-1.02	Zn		Ist(b)$^-$			-2e			Zn(Ist(b))$^+$	
-1.02	Zn		Oxd$^-$			-2e			Zn(Oxd)$^+$	
-1.02	Zn		2Oxin$^-$			-2e			Zn(Oxin)$_2$	
-1.01	Zn		Himda^{-2}			-2e			Zn(Himda)	
-1.01	Zn		Met(a)$^-$			-2e			Zn(Met(a))$^+$	
-1.01	Zn		Met(b)$^-$			-2e			Zn(Met(b))$^+$	
-1.00	Zn		2meOxin$^-$			-2e			Zn(meOxin)$_2$	
-1.00	Zn		Ndpa^{-3}			-2e			Zn(Ndpa)$^-$	
-1.00	Zn		2Oxd$^-$			-2e			Zn(Oxd)$_2$	
-0.99	Zn		DmHas$^-$			-2e			Zn(DmHas)$^+$	
-0.99	Zn		2DmHas$^-$			-2e			Zn(DmHas)$_2$	
-0.98	Zn		Coy$^-$			-2e			Zn(Coy)$^+$	
-0.98	Zn		Has$^-$			-2e			Zn(Has)$^+$	
-0.98	Zn		2Ist(a)$^-$			-2e			Zn(Ist(a))$_2$	
-0.98	Zn		Mcin$^-$			-2e			Zn(Mcin)$^+$	

EMF	Elem	Anion	Compound or anion	Cation	H₂O	e =	Cation	Anion	Complex or compound	Compound or element
-0.98	Zn		Trop⁻			-2e			Zn(Trop)⁺	
-0.97	Zn		Imda⁻²			-2e			Zn(Imda)	
-0.97	Zn		2Met(a)⁻			-2e			Zn(Met(a))₂	
-0.97	Zn		2Ist(b)⁻			-2e			Zn(Ist(b))₂	
-0.97	Zn		MpHas⁻			-2e			Zn(MpHas)⁺	
-0.97	Zn		Hox⁻			-2e			Zn(Hox)⁺	
-0.97	Zn		2Has⁻			-2e			Zn(Has)₂	
-0.97	Zn		Cin⁻			-2e			Zn(Cin)⁺	
-0.96	Zn		2Met(b)⁻			-2e			Zn(Met(b))₂	
-0.96	Zn		Ptn			-2e			Zn(Ptn)⁺²	
-0.95	Zn		Af⁻			-2e			Zn(Af)⁺	
-0.95	Zn		2Mcin⁻			-2e			Zn(Mcin)₂	
-0.95	Zn		Ph			-2e			Zn(Ph)⁺²	
-0.95	Zn		2Trop⁻			-2e			Zn(Trop)₂	
-0.94	Zn		Impa⁻²			-2e			Zn(Impa)₂	
-0.93	Zn		2Af⁻			-2e			Zn(Af)₂	
-0.93	Zn		Aspa⁻²			-2e			Zn(Aspa)	
-0.93	Zn		2Cin⁻			-2e			Zn(Cin)₂	
-0.93	Zn		2Coy⁻			-2e			Zn(Coy)₂	
-0.93	Zn		En			-2e			Zn(En)⁺²	
-0.93	Zn		2Hox⁻			-2e			Zn(Hox)₂	
-0.93	Zn		2MpHas⁻			-2e			Zn(MpHas)₂	
-0.93	Zn		2Ph			-2e			Zn(Ph)₂⁺²	
-0.93	Zn		Pn			-2e			Zn(Pn)⁺²	
-0.92	Zn		2Deta			-2e			Zn(Deta)₂⁺²	
-0.92	Zn		Gl⁻			-2e			Zn(Gl)⁺	
-0.92	Zn		Dge⁻			-2e			Zn(Dge)⁺	
-0.92	Zn		Ntp⁻³			-2e			Zn(Ntp)⁻	
-0.91	Zn		AcAc⁻			-2e			Zn(AcAc)⁺	
-0.91	Zn		Alan⁻			-2e			Zn(Alan)⁺	
-0.91	Zn		2Imda⁻²			-2e			Zn(Imda)₂⁻²	
-0.91	Zn		Imdp⁻²			-2e			Zn(Imdp)	

EMF	Elem	Anion	Compound or anion	Cation	H_2O	e =	Cation	Anion	Complex or compound	Compound or element
-0.91	Zn		mOxin$^-$			-2e			Zn(mOxin)$^+$	
-0.91	Zn		2Pn			-2e			Zn(Pn)$_2^{+2}$	
-0.90	Zn		2En			-2e			Zn(En)$_2^{+2}$	
-0.90	Zn		Ox^{-2}			-2e			Zn(Ox)	
-0.90	Zn		3Ph			-2e			Zn(Ph)$_3^{+2}$	
-0.89	Zn		2Alan$^-$			-2e			Zn(Alan)$_2$	
-0.89	Zn		2Aspa^{-2}			-2e			Zn(Aspa)$_2^{-2}$	
-0.89	Zn		2Gl$^-$			-2e			Zn(Gl)$_2$	
-0.89	Zn		2Impa^{-2}			-2e			Zn(Impa)$_2^{-2}$	
-0.89	Zn		2mOxin$^-$			-2e			Zn(mOxin)$_2$	
-0.89	Zn		OH$^-$			-2e			ZnOH$^+$	
-0.89	Zn		Sald$^-$			-2e			Zn(Sald)$^+$	
-0.88	Zn		2AcAc$^-$			-2e			Zn(AcAc)$_2$	
-0.88	Zn		2Himda^{-2}			-2e			Zn(Himda)$_2^{-2}$	
-0.87	Zn		Glgl$^-$			-2e			Zn(Glgl)$^+$	
-0.87	Zn		3Ist(b)$^-$			-2e			Zn(Ist(b))$_3^-$	
-0.87	Zn		3Met(b)$^-$			-2e			Zn(Met(b))$_3^-$	
-0.87	Zn		2Sald$^=$			-2e			Zn(Sald)$_2$	
-0.86	Zn		2Dge$^-$			-2e			Zn(Dge)$_2$	
-0.86	Zn		Mal^{-2}			-2e			Zn(Mal)	
-0.86	Zn		3Trop$^-$			-2e			Zn(Trop)$_3^-$	
-0.85	Zn		2Amac^{-3}			-2e			Zn(Amac)$_2^{-4}$	
-0.85	Zn		2Nta^{-3}			-2e			Zn(Nta)$_2^{-4}$	
-0.85	Zn		SSald^{-2}			-2e			Zn(SSald)	
-0.84	Zn		Ap^{-2}			-2e			Zn(Ap)	
-0.84	Zn		2Glgl$^-$			-2e			Zn(Glgl)$_2$	
-0.84	Zn		Im			-2e			Zn(Im)$^{+2}$	
-0.84	Zn		Tart^{-2}			-2e			Zn(Tart)	
-0.83	Zn		2Im			-2e			Zn(Im)$_2^{+2}$	
-0.83	Zn		3Im			-2e			Zn(Im)$_3^{+2}$	
-0.83	Zn		NH_3			-2e			Zn(NH_3)$^{+2}$	
-0.83	Zn		2NH_3			-2e			Zn(NH_3)$_2^{+2}$	

EMF	Elem	Anion	Compound or anion	Cation	H_2O	e =	Cation	Anion	Complex or compound	Compound or element
-0.83	Zn		$3NH_3$			-2e			$Zn(NH_3)_3^{+2}$	
-0.83	Zn		N_2H_4			-2e			$Zn(N_2H_4)^{+2}$	
-0.83	Zn		SO_4^{-2}			-2e			$ZnSO_4$	
-0.83	Zn		$S_2O_3^{-2}$			-2e			ZnS_2O_3	
-0.82	Zn		Gly^-			-2e			$Zn(Gly)^+$	
-0.82	Zn		$4Im$			-2e			$Zn(Im)_4^{+2}$	
-0.82	Zn		Lac^-			-2e			$Zn(Lac)^{+2}$	
-0.82	Zn		$4NH_3$			-2e			$Zn(NH_3)_4^{+2}$	
-0.81	Zn		Ac^-			-2e			$Zn(Ac)^+$	
-0.81	Zn		CNS^-			-2e			$ZnCNS^+$	
-0.81	Zn		$3En$			-2e			$Zn(En)_3^{+2}$	
-0.81	Zn		$Glac^-$			-2e			$Zn(Glac)^+$	
-0.81	Zn		Glu^-			-2e			$Zn(Glu)^+$	
-0.81	Zn		$2N_2H_4$			-2e			$Zn(N_2H_4)_2^{+2}$	
-0.81	Zn		$3Pn$			-2e			$Zn(Pn)_3^{+2}$	
-0.81	Zn		Suc^{-2}			-2e			$Zn(Suc)$	
-0.80	Zn		$3N_2H_4$			-2e			$Zn(N_2H_4)_3^{+2}$	
-0.80	Zn		F^-			-2e			ZnF^+	
-0.79	Zn		Bu^-			-2e			$Zn(Bu)^+$	
-0.79	Zn		Pr^-			-2e			$Zn(Pr)^+$	
-0.78	Zn		$3Cl^-$			-2e			$ZnCl_3^-$	
-0.78	Zn		$4N_2H_4$			-2e			$Zn(N_2H_4)_4^{+2}$	
-0.76	Zn					-2e	Zn^{+2}			
-0.76			$Zn(Hg)$			-2e	Zn^{+2}			Hg
-0.76	Zn		Nac^-			-2e			$Zn(Nac)^+$	
-0.75	Zn		Br^-			-2e			$ZnBr^+$	
-0.75	Zn		Cl^-			-2e			$ZnCl^+$	
-0.75	Zn		$2Cl^-$			-2e			$ZnCl_2$	
-0.72	Zn		I^-			-2e			ZnI^+	
-0.44	Zn				H_2O	-2e	$2H^+$		ZnO	
+0.05	Zn				$2H_2O$	-2e	$3H^+$	$HZnO_2^-$		

EMF	Elem	Anion	Compound or anion	Cation	H_2O	e =	Cation	Anion	Complex or compound	Compound or element
-2.36	Zr		$4OH^-$			-4e			H_2ZrO_3	H_2O
-1.66	Zr		F^-			-4e			ZrF^{+3}	
-1.64	Zr		$2F^-$			-4e			ZrF_2^{+2}	
-1.62	Zr		$3F^-$			-4e			ZrF_3^+	
-1.59	Zr		SO_4^{-2}			-4e			$ZrSO_4^{+2}$	
-1.57	Zr		$2SO_4^{-2}$			-4e			$Zr(SO_4)_2$	
-1.55	Zr		$3SO_4^{-2}$			-4e			$Zr(SO_4)_3^{-2}$	
-1.53	Zr					-4e	Zr^{+4}			
-1.55	Zr				$4H_2O$	-4e	$4H^+$		$Zr(OH)_4$	
-1.46	Zr				$2H_2O$	-4e	$4H^+$		ZrO_2	

PART III

CHEMICAL ELECTRODE POTENTIALS

OF

INORGANIC ANIONS

EMF	Anion	Elem	Compound or anion	Cation	H_2O	e =	Cation	Anion	Complex or compound	Compound or element
-1.91	BrO_3^-	Th				-4e			$ThBrO_3^{+3}$	
-1.90	$2BrO_3^-$	Th				-4e			$Th(BrO_3)_2^{+2}$	

EMF	Anion	Elem	Compound or anion	Cation	H₂O	e =	Cation	Anion	Complex or compound	Compound or element
-2.49	Br⁻	Ce				-3e			$CeBr^{+2}$	
-2.40	Br⁻	U				-4e			UBr^{+3}	
-0.75	Br⁻	Zn				-2e			$ZnBr^{+}$	
-0.66	Br⁻	Tl				-e			$TlBr$	
-0.45	Br⁻	Cd				-2e			$CdBr^{+}$	
-0.43	3Br⁻	Cd				-2e			$CdBr_3^{-}$	
-0.42	2Br⁻	Cd				-2e			$CdBr_2$	
-0.41	4Br⁻	Cd				-2e			$CdBr_4^{-2}$	
-0.36	Br⁻	In				-3e			$InBr^{+2}$	
-0.35	2Br⁻	In				-3e			$InBr_2^{+}$	
-0.35	3Br⁻	In				-3e			$InBr_3$	
-0.28	2Br⁻	Pb				-2e			$PbBr_2$	
-0.16	Br⁻	Pb				-2e			$PbBr^{+}$	
-0.16	Br⁻	Sn				-2e			$SnBr^{+}$	
-0.15	2Br⁻	Sn				-2e			$SnBr_2$	
-0.15	3Br⁻	Sn				-2e			$SnBr_3^{-}$	
+0.03	Br⁻	Cu				-e			$CuBr$	
+0.07	Br⁻	Ag				-e			$AgBr$	
+0.07	Br⁻			UO_2^{+}		-e			UO_2Br^{+}	
+0.12	Br⁻	Bi				-3e			$BiBr^{+2}$	
+0.14	2Br⁻	2Hg				-2e			Hg_2Br_2	
+0.17	2Br⁻	Cu				-e			$CuBr_2^{-}$	
+0.18	2Br⁻	Bi				-3e			$BiBr_2^{+}$	
+0.19	3Br⁻	Bi				-3e			$BiBr_3$	
+0.22	4Br⁻	Hg				-2e			$HgBr_4^{-2}$	
+0.28	2Br⁻	Cu				-2e			$CuBr_2$	
+0.33	Br⁻	Cu				-2e			$CuBr^{+}$	
+0.34	Br⁻	Fe				-3e			$FeBr^{+2}$	
+0.55	Br⁻	Ag				-e			$AgBr$	
+0.58	4Br⁻	Pt				-2e			$PtBr_4^{-2}$	
+0.62	2Br⁻	Ag				-e			$AgBr_2^{-}$	
+0.69	3Br⁻	Ag				-e			$AgBr_3^{-2}$	

EMF	Anion	Elem	Compound or anion	Cation	H_2O	e =	Cation	Anion	Complex or compound	Compound or element
+0.72	Br^-	Tl				-3e			$TlBr^{+2}$	
+0.78	$4Br^-$	Ag				-e			$AgBr_4^{-3}$	
+0.82	$3Br^-$	Tl				-3e			$TlBr_3$	
+0.86	$4Br^-$	Tl				-3e			$TlBr_4^-$	
+0.87	$4Br^-$	Au				-3e			$AuBr_4^-$	
+0.96	$2Br^-$	Au				-e			$AuBr_2^-$	

EMF	Anion	Elem	Compound or anion	Cation	H_2O	e =	Cation	Anion	Complex or compound	Compound or element
-2.92	ClO_3^-	Ba				-2e			$BaClO_3^+$	
-1.90	ClO_3^-	Th				-4e			$ThClO_3^{+3}$	

EMF	Anion	Elem	Compound or anion	Cation	H_2O	e = Cation	Anion	Complex or compound	Compound or element
-2.41	Cl^-	U				-4e		UCl^{+3}	
-1.90	Cl^-	Th				-4e		$ThCl^{+3}$	
-1.89	$3Cl^-$	Th				-4e		$ThCl_3^+$	
-1.89	$4Cl^-$	Th				-4e		$ThCl_4$	
-1.88	$2Cl^-$	Th				-4e		$ThCl_2^{+2}$	
-0.78	$3Cl^-$	Zn				-2e		$ZnCl_3^-$	
-0.75	Cl^-	Zn				-2e		$ZnCl^+$	
-0.75	$2Cl^-$	Zn				-2e		$ZnCl_2$	
-0.55	Cl^-	Ga				-3e		$GaCl^{+2}$	
-0.51	$2Cl^-$	Ga				-3e		$GaCl_2^+$	
-0.47	$3Cl^-$	Ga				-3e		$GaCl_3$	
-0.45	$4Cl^-$	Ga				-3e		$GaCl_4^-$	
-0.42	$2Cl^-$	Cd				-2e		$CdCl_2$	
-0.41	$3Cl^-$	Cd				-2e		$CdCl_3^-$	
-0.38	Cl^-	Tl				-e		$TlCl$	
-0.37	Cl^-	In				-3e		$InCl^{+2}$	
-0.36	$2Cl^-$	In				-3e		$InCl_2^+$	
-0.36	$3Cl^-$	In				-3e		$InCl_3$	
-0.27	$2Cl^-$	Pb				-2e		$PbCl_2$	
-0.18	Cl^-	Pb				-2e		$PbCl^+$	
-0.17	Cl^-	Sn				-2e		$SnCl^+$	
-0.16	$2Cl^-$	Sn				-2e		$SnCl_2$	
-0.14	$3Cl^-$	Sn				-2e		$SnCl_3^-$	
+0.07	Cl^-			UO_2^+		-e		UO_2Cl^+	
+0.15	Cl^-	Bi				-3e		$BiCl^{+2}$	
+0.16	Cl^-	Bi			H_2O	-3e	$2H^+$	$BiOCl$	
+0.19	$2Cl^-$	Bi				-3e		$BiCl_2^+$	
+0.19	$3Cl^-$	Bi				-3e		$BiCl_3$	
+0.20	$4Cl^-$	Bi				-3e		$BiCl_4^-$	
+0.27	$2Cl^-$	2Hg				-2e		Hg_2Cl_2	

EMF	Anion	Elem	Compound or anion	Cation	H_2O	e =	Cation	Anion	Complex or compound	Compound or element
+0.31	Cl^-	Mn				-3e			$MnCl^{+2}$	
+0.32	Cl^-	Fe				-3e			$FeCl^{+2}$	
+0.33	$2Cl^-$	Fe				-3e			$FeCl_2^+$	
+0.36	$3Cl^-$	Fe				-3e			$FeCl_3$	
+0.64	Cl^-	Ag				-e			$AgCl$	
+0.68	$2Cl^-$	Ag				-e			$AgCl_2^-$	
+0.68	$2Cl^-$		$PtCl_4^{-2}$			-2e			$PtCl_6^{-2}$	
+0.75	Cl^-	Tl				-3e			$TlCl^{+2}$	
+0.80	$4Cl^-$	Ag				-e			$AgCl_4^{-3}$	
+0.80	$2Cl^-$	Tl				-3e			$TlCl_2^+$	
+0.86	$4Cl^-$	Tl				-3e			$TlCl_4^-$	
+0.87	$3Cl^-$	Tl				-3e			$TlCl_3$	
+1.00	$4Cl^-$	Au				-3e			$AuCl_4^-$	
+1.15	$2Cl^-$	Au				-e			$AuCl_2^-$	

EMF	Anion	Elem	Compound or anion	Cation	H_2O	e =	Cation	Anion	Complex or compound	Compound or element
-1.28	$4CN^-$	Cu				-e			$Cu(CN)_4^{-3}$	
-1.26	$4CN^-$	Zn				-2e			$Zn(CN)_4^{-2}$	
-1.17	$3CN^-$	Cu				-e			$Cu(CN)_3^{-2}$	
-1.03	$4CN^-$	Cd				-2e			$Cd(CN)_4^{-2}$	
-0.64	CN^-	Cu				-e			CuCN	
-0.56	CN^-	Cd				-2e			$CuCN^+$	
-0.55	$2CN^-$	Cd				-2e			$Cd(CN)_2$	
-0.54	$3CN^-$	Cd				-2e			$Cd(CN)_3^-$	
-0.51	$4CN^-$	Cd				-2e			$Cd(CN)_4^{-2}$	
-0.43	$2CN^-$	Cu				-e			$Cu(CN)_2^-$	
-0.37	$4CN^-$	Hg				-2e			$Hg(CN)_4^{-2}$	
-0.31	$2CN^-$	Ag				-e			$Ag(CN)_2^-$	
-0.02	CN^-	Ag				-e			AgCN	
+0.74	$3CN^-$	Ag				-e			$Ag(CN)_3^{-2}.$	
+0.83	$4CN^-$	Ag				-e			$Ag(CN)_4^{-3}$	

EMF	Anion	Elem	Compound or anion	Cation	H_2O	e =	Cation	Anion	Complex or compound	Compound or element
-2.57	F^-	La				-3e			LaF^{+2}	
-2.54	F^-	Ce				-3e			CeF^{+2}	
-2.45	F^-	Gd				-3e			GdF^{+2}	
-2.43	$3F^-$	Sc				-3e			ScF_3	
-2.38	F^-	Mg				-2e			MgF^+	
-2.33	$2F^-$	Sc				-3e			ScF_2^+	
-2.24	F^-	Sc				-3e			ScF^{+2}	
-2.07	$6F^-$	Al				-3e			AlF_6^{-3}	
-2.01	F^-	Th				-4e			ThF^{+3}	
-1.99	$2F^-$	Th				-4e			ThF_2^{+2}	
-1.97	$3F^-$	Th				-4e			ThF_3^+	
-1.83	F^-	Be				-2e			BeF^+	
-1.79	F^-	Al				-3e			AlF^{+2}	
-1.77	$2F^-$	Al				-3e			AlF_2^+	
-1.76	$2F^-$	Be				-2e			BeF_2	
-1.75	$3F^-$	Al				-3e			AlF_3	
-1.72	$4F^-$	Al				-3e			AlF_4^-	
-1.70	$5F^-$	Al				-3e			AlF_5^{-2}	
-1.68	$6F^-$	Al				-3e			AlF_6^{-3}	
-1.66	F^-	Zr				-4e			ZrF^{+3}	
-1.64	$2F^-$	Zr				-4e			ZrF_2^{+2}	
-1.62	$3F^-$	Zr				-4e			ZrF_3^+	
-1.19	$6F^-$	Ti				-4e			TiF_6^{-2}	
-1.16	F^-	Pu				-4e			PuF^{+3}	
-0.83	F^-	Cr				-3e			CrF^{+2}	
-0.81	$2F^-$	Cr				-3e			CrF_2^+	
-0.79	$3F^-$	Cr				-3e			CrF_3	
-0.63	F^-	Ga				-3e			GaF^{+2}	
-0.41	F^-	In				-3e			InF^{+2}	
-0.39	$2F^-$	In				-3e			InF_2^+	
-0.39	$3F^-$	In				-3e			InF_3	
-0.36	$4F^-$	In				-3e			InF_4^-	

EMF	Anion	Elem	Compound or anion	Cation	H_2O	e =	Cation	Anion	Complex or compound	Compound or element
-0.35	F^-	Tl				-e				TlF
-0.22	F^-			UO_2^+		-e				UO_2F^+
-0.19	F^-	H				-e				HF
-0.15	$2F^-$			UO_2^+		-e				UO_2F_2
-0.10	$3F^-$			UO_2^+		-e				$UO_2F_3^-$
-0.03	$4F^-$			UO_2^+		-e				$UO_2F_4^{-2}$
+0.23	F^-	Fe				-3e				FeF^{+2}
+0.25	$2F^-$	Fe				-3e				FeF_2^+
+0.28	$3F^-$	Fe				-3e				FeF_3

EMF	Anion	Elem	Compound or anion	Cation	H_2O	e =	Cation	Anion	Complex or compound	Compound or element
-3.03	OH^-	Li				-e			LiOH	
-3.02	$2OH^-$	Ca				-2e			$Ca(OH)_2$	
-2.99	$2OH^-$	Ba			$8H_2O$	-2e			$Ba(OH)_2 \cdot 8H_2O$	
-2.92	OH^-	Ba				-2e			$BaOH^+$	
-2.91	OH^-	Sr				-2e			$SrOH^+$	
-2.91	OH^-	Ca				-2e			$CaOH^+$	
-2.90	$3OH^-$	La				-3e			$La(OH)_3$	
-2.88	$2OH^-$	Sr				-2e				$Sr(OH)_2$
-2.87	$3OH^-$	Ce				-3e			$Ce(OH)_3$	
-2.77	OH^-	Ce				-3e			$CeOH^{+2}$	
-2.69	$2OH^-$	Mg				-2e			$Mg(OH)_2$	
-2.68	OH^-	Na				-e			NaOH	
-2.63	$6OH^-$	2Be				-4e		$Be_2O_3^-$		$3H_2O$
-2.62	OH^-	La				-3e			$LaOH^{+2}$	
-2.61	$2OH^-$	Be				-2e			H_2O	BeO
-2.61	$3OH^-$	Sc				-3e			$Sc(OH)_3$	
-2.58	OH^-	U				-4e			UOH^{+3}	
-2.48	$4OH^-$	Th				-4e			$Th(OH)_4$	
-2.42	OH^-	Mg				-2e			$MgOH^+$	
-2.42	$3OH^-$	Pu				-3e			$Pu(OH)_3$	
-2.39	$4OH^-$	U				-4e			$2H_2O$	UO_2
-2.33	$4OH^-$	Al				-3e		$H_2AlO_3^-$		H_2O
-2.32	OH^-	Sc				-3e			$ScOH^{+2}$	
-2.30	$3OH^-$	Al				-3e			$Al(OH)_3$	
-2.17	$3OH^-$	U				-3e			$U(OH)_3$	
-1.92	OH^-	Be				-2e			$BeOH^+$	
-1.91	OH^-	Th				-4e			$ThOH^{+3}$	
-1.84	OH^-	Al				-3e			$AlOH^{+2}$	
-1.67	OH^-	V				-3e			VOH^{+2}	
-1.66	$2OH^-$	V				-3e			$V(OH)_2^+$	
-1.55	$2OH^-$	Mn				-2e			$Mn(OH)_2$	
-1.48	$3OH^-$	Cr				-3e			$Cr(OH)_3(c)$	

EMF	Anion	Elem	Compound or anion	Cation	H_2O	e =	Cation	Anion	Complex or compound	Compound or element
-1.34	$30H^-$	Cr				-3e			$Cr(OH)_3(hyd)$	
-1.27	$40H^-$	Cr				-3e			CrO_2^-	$2H_2O$
-1.25	$20H^-$	Zn				-2e			$Zn(OH)_2$	
-1.22	$40H^-$	Ga				-3e		$H_2GaO_3^-$	H_2O	
-1.22	$40H^-$	Zn				-2e		ZnO_2^{-2}	$2H_2O$	
-1.19	OH^-	Pu				-4e			$PuOH^{+3}$	
-1.17	OH^-	Tl				-e			$TlOH$	
-1.15	OH^-	Mn				-2e			$MnOH^+$	
-1.15	$330H^-$	6V				-30e		$HV_6O_{17}^{-3}$	$16H_2O$	
-1.00	$30H^-$	In				-3e			$In(OH)_3$	
-0.94	OH^-	Cr				-3e			$CrOH^{+2}$	
-0.93	$30H^-$		$HSnO_2^-$		H_2O	-2e		$Sn(OH)_6^-$		
-0.91	$30H^-$	Sn				-2e		$HSnO_2^-$	H_2O	
-0.89	OH^-	Zn				-2e			$ZnOH^+$	
-0.88	$20H^-$	Fe				-2e			$Fe(OH)_2$	
-0.81	$20H^-$	Cd				-2e			$Cd(OH)_2$	
-0.74	OH^-	Ga				-3e			$GaOH^{+2}$	
-0.73	$20H^-$	Co				-2e			$Co(OH)_2$	
-0.72	$20H^-$	Ni				-2e			$Ni(OH)_2$	
-0.58	$20H^-$	Pb				-2e			$PbO (r)$	H_2O
-0.56	OH^-	Fe				-2e			$FeOH^+$	
-0.54	OH^-	In				-3e			$InOH^{+2}$	
-0.54	$30H^-$	Pb				-2e		$HPbO_2^-$	H_2O	
-0.47	OH^-	Cd				-2e			$CdOH^+$	
-0.43	OH^-	Sn				-2e			$SnOH^+$	
-0.41	OH^-	Co				-2e			$CoOH^+$	
-0.39	OH^-	Ni				-2e			$NiOH^+$	
-0.36	$20H^-$	2Cu				-2e			Cu_2O	H_2O
-0.34	OH^-	Tl				-e			$TlOH (c)$	
-0.31	OH^-	Pb				-2e			$PbOH^+$	
-0.26	$20H^-$	Cu				-2e			CuO	H_2O
-0.22	$20H^-$	Cu				-2e			$Cu(OH)_2$	

EMF	Anion	Elem	Compound or anion	Cation	H_2O	e =	Cation	Anion	Complex or compound	Compound or element
+0.10	OH^-	Fe				-3e			$FeOH^{+2}$	
+0.10	$2OH^-$	Hg				-2e			HgO (r)	H_2O
+0.12	$2OH^-$	Fe				-3e			$Fe(OH)_2^+$	
+0.15	OH^-	Cu				-2e			$CuOH^+$	
+0.35	$2OH^-$	2Ag				-2e			Ag_2O	H_2O
+0.63	OH^-	Tl				-3e			$TlOH^{+2}$	
+0.65	$2OH^-$	Tl				-3e			$Tl(OH)_2^+$	
+0.66	OH^-	Ag				-e			$AgOH$	

EMF	Anion	Elem	Compound or anion	Cation	H_2O	e =	Cation	Anion	Complex or compound	Compound or element
-2.93	IO_3^-	Ba				-2e			$BaIO_3^+$	
-2.92	IO_3^-	Sr				-2e			$SrIO_3^+$	
-2.90	IO_3^-	Ca				-2e			$CaIO_3^+$	
-2.90	IO_3^-	K				-e			KIO_3	
-2.36	IO_3^-	Mg				-2e			$MgIO_3^+$	
-1.94	IO_3^-	Th				-4e			$ThIO_3^{+3}$	
-0.37	IO_3^-	Tl				-e			$TlIO_3$	

EMF	Anion	Elem	Compound or anion	Cation	H_2O	e =	Cation	Anion	Complex or compound	Compound or element
-0.72	I^-	Zn				-2e			ZnI^+	
-0.47	I^-	Cd				-2e			CdI^+	
-0.47	$3I^-$	Cd				-2e			CdI_3^-	
-0.45	$2I^-$	Cd				-2e			CdI_2	
-0.44	$4I^-$	Cd				-2e			CdI_4^{-2}	
-0.35	I^-	In				-3e			InI^{+2}	
-0.19	I^-	Cu				-e			CuI	
-0.15	I^-	Ag				-e			AgI	
-0.04	$2I^-$	2Hg				-2e			Hg_2I_2	
-0.04	$4I^-$	Hg				-2e			HgI_4^{-2}	

EMF	Anion	Elem	Compound or anion	Cation	H_2O	e =	Cation	Anion	Complex or compound	Compound or element
-2.93	NO_3^-	Ba				-2e			$BaNO_3^+$	
-2.91	NO_3^-	Sr				-2e			$SrNO_3^+$	
-2.88	NO_3^-	Ca				-2e			$CaNO_3^+$	
-2.34	NO_3^-	Mg				-2e			$MgNO_3^+$	
-1.91	NO_3^-	Th				-4e			$ThNO_3^{+3}$	
-1.07	NO_3^-	Pu				-4e			$PuNO_3^{+3}$	
-0.41	NO_3^-	Cd				-2e			$CdNO_3^+$	
-0.14	NO_3^-	Pb				-2e			$PbNO_3^+$	
+0.09	NO_3^-			UO_2^+		-e			$UO_2NO_3^+$	
+0.16	NO_3^-	Bi				-3e			$BiNO_3^{+2}$	
+0.91	NO_3^-	Tl				-3e			$TlNO_3^{+2}$	

EMF	Anion	Elem	Compound or anion	Cation	H_2O	e =	Cation	Anion	Complex or compound	Compound or element
-2.94	HPO_4^{-2}	Ca				-2e				$CaHPO_4$
-2.44	PO_4^{-3}	Mg				-2e				$MgPO_4^-$
-2.41	HPO_4^{-2}	Mg				-2e				$MgHPO_4$
-1.96	$H_2PO_4^-$	Th				-4e				$ThH_2PO_4^{+3}$
-1.96	$2H_2PO_4^-$	Th				-4e				$Th(H_2PO_4)_2^{+2}$
-1.93	H_3PO_4	Th				-4e				$Th(H_3PO_4)^{+4}$

EMF	Anion	Elem	Compound or anion	Cation	H_2O	e =	Cation	Anion	Complex or compound	Compound or element
-3.02	$P_2O_7^{-4}$	Ca				-2e			$CaP_2O_7^{-2}$	
-2.82	$P_2O_7^{-4}$	Ce				-3e			$CeP_2O_7^{-}$	
-2.51	$P_2O_7^{-4}$	Mg				-2e			$MgP_2O_7^{-2}$	
-0.44	$P_2O_7^{-4}$	Tl				-e			$TlP_2O_7^{-3}$	
-0.42	$P_2O_7^{-4}$	Ni				-2e			$NiP_2O_7^{-2}$	
-0.35	$2P_2O_7^{-4}$	Tl				-e			$Tl(P_2O_7)_2^{-7}$	
-0.29	$2P_2O_7^{-4}$	Ni				-2e			$Ni(P_2O_7)_2^{-6}$	
+0.14	$P_2O_7^{-4}$	Cu				-2e			$CuP_2O_7^{-2}$	

EMF	Anion	Elem	Compound or anion	Cation	H_2O	e =	Cation	Anion	Complex or compound	Compound or element
+0.03	4SeCN$^-$	Hg				-2e			Hg(SeCN)$_4^{-2}$	
+0.73	3SeCN$^-$	Ag				-e			Ag(SeCN)$_3^{-2}$	

EMF	Anion	Elem	Compound or anion	Cation	H_2O	e =	Cation	Anion	Complex or compound	Compound or element
-2.98	SO_4^{-2}	K				-e			KSO_4^{-}	
-2.94	SO_4^{-2}	Ca				-2e			$CaSO_4$	
-2.75	SO_4^{-2}	Na				-e			$NaSO_4^{-}$	
-2.59	SO_4^{-2}	La				-3e			$LaSO_4^{+}$	
-2.53	SO_4^{-2}	Pr				-3e			$PrSO_4^{+}$	
-2.52	SO_4^{-2}	Ce				-3e			$CeSO_4^{+}$	
-2.50	SO_4^{-2}	Nd				-3e			$NdSO_4^{+}$	
-2.48	SO_4^{-2}	Sm				-3e			$SmSO_4^{+}$	
-2.47	SO_4^{-2}	Gd				-3e			$GdSO_4^{+}$	
-2.45	SO_4^{-2}	U				-4e			USO_4^{+2}	
-2.44	SO_4^{-2}	Y				-3e			YSO_4^{+}	
-2.43	$2SO_4^{-2}$	U				-4e			$U(SO_4)_2$	
-2.41	SO_4^{-2}	Mg				-2e			$MgSO_4$	
-2.39	SO_4^{-2}	Ho				-3e			$HoSO_4^{+}$	
-2.37	SO_4^{-2}	Er				-3e			$ErSO_4^{+}$	
-2.34	SO_4^{-2}	Yb				-3e			$YbSO_4^{+}$	
-1.95	SO_4^{-2}	Th				-4e			$ThSO_4^{+2}$	
-1.93	$2SO_4^{-2}$	Th				-4e			$Th(SO_4)_2$	
-1.75	SO_4^{-2}	Np				-4e			$NpSO_4^{+2}$	
-1.59	SO_4^{-2}	Zr				-4e			$ZrSO_4^{+2}$	
-1.57	$2SO_4^{-2}$	Zr				-4e			$Zr(SO_4)_2$	
-1.55	$3SO_4^{-2}$	Zr				-4e			$Zr(SO_4)_3^{-2}$	
-1.12	SO_4^{-2}	Mn				-2e			$MnSO_4$	
-1.11	SO_4^{-2}	Pu				-4e			$PuSO_4^{+2}$	
-0.83	SO_4^{-2}	Zn				-2e			$ZnSO_4$	
-0.51	SO_4^{-2}	Fe				-2e			$FeSO_4$	
-0.47	SO_4^{-2}	Cd				-2e			$CdSO_4$	
-0.38	SO_4^{-2}	In				-3e			$InSO_4^{+}$	
-0.35	$3SO_4^{-2}$	In				-3e			$In(SO_4)_3^{-3}$	
-0.35	SO_4^{-2}	Co				-2e			$CoSO_4$	
-0.34	$2SO_4^{-2}$	In				-3e			$In(SO_4)_2^{-}$	
-0.32	SO_4^{-2}	Ni				-2e			$NiSO_4$	

EMF	Anion	Elem	Compound or anion	Cation	H_2O	e =	Cation	Anion	Complex or compound	Compound or element
-0.07	SO_4^{-2}	H				-e			HSO_4^-	
-0.06	SO_4^{-2}	2H				-2e			H_2SO_4	
-0.05	SO_4^{-2}			UO_2^+		-e			UO_2SO_4	
0.00	$3SO_4^{-2}$			UO_2^+		-e			$UO_2(SO_4)_3^{-4}$	
+0.01	$2SO_4^{-2}$			UO_2^+		-e			$UO_2(SO_4)_2^{-2}$	
+0.27	SO_4^{-2}	Cu				-2e			$CuSO_4^+$	
+0.27	SO_4^{-2}	Fe				-2e			$FeSO_4^+$	
+0.31	$2SO_4^{-2}$	Fe				-3e			$Fe(SO_4)_2^-$	
+0.65	SO_4^{-2}	2Ag				-2e			Ag_2SO_4	
+0.79	SO_4^{-2}	Ag				-e			$AgSO_4^-$	
+0.80	$2SO_4^{-2}$	Ag				-e			$Ag(SO_4)_2^{-3}$	

EMF	Anion	Elem	Compound or anion	Cation	H_2O	e =	Cation	Anion	Complex or compound	Compound or element
+0.08	SO_3^{-2}	Cu				-e			$CuSO_3^-$	
+0.46	$2SO_3^{-2}$	Cu				-e			$Cu(SO_3)_2^{-3}$	
+0.48	$3SO_3^{-2}$	Cu				-e			$Cu(SO_3)_3^{-5}$	
+0.49	SO_3^{-2}	Ag				-e			$AgSO_3^-$	
+0.68	$2SO_3^{-2}$	Ag				-e			$Ag(SO_3)_2^{-3}$	

EMF	Anion	Elem	Compound or anion	Cation	H_2O	e =	Cation	Anion	Complex or compound	Compound or element
-3.05	$P_4O_{12}^{-4}$	Ba				-2e			$BaP_4O_{12}^{-2}$	
-3.04	$P_4O_{12}^{-4}$	Sr				-2e			$SrP_4O_{12}^{-2}$	
-3.03	$P_4O_{12}^{-4}$	Ca				-2e			$CaP_4O_{12}^{-2}$	
-2.76	$P_4O_{12}^{-4}$	Na				-e			$NaP_4O_{12}^{-3}$	
-2.65	$P_4O_{12}^{-4}$	La				-3e			$LaP_4O_{12}^{-}$	
-2.49	$P_4O_{12}^{-4}$	Mg				-2e			$MgP_4O_{12}^{-2}$	
-1.21	$P_4O_{12}^{-4}$	Mn				-2e			$MnP_4O_{12}^{-2}$	
-0.40	$P_4O_{12}^{-4}$	Ni				-2e			$NiP_4O_{12}^{-2}$	
-0.28	$2P_4O_{12}^{-4}$	Ni				-2e			$Ni(P_4O_{12})_2^{-6}$	
+0.25	$P_4O_{12}^{-4}$	Cu				-2e			$CuP_4O_{12}^{-2}$	
+0.30	$2P_4O_{12}^{-4}$	Cu				-2e			$Cu(P_4O_{12})_2^{-6}$	

EMF	Anion	Elem	Compound or anion	Cation	H_2O	e =	Cation	Anion	Complex or compound	Compound or element
-2.42	CNS^-	U				-4e			$UCNS^{+3}$	
-2.41	$2CNS^-$	U				-4e			$U(CNS)_2^{+2}$	
-1.49	CNS^-	V				-3e			$VCNS^{+2}$	
-0.81	CNS^-	Zn				-2e			$ZnCNS^+$	
-0.78	CNS^-	Cr				-3e			$CrCNS^{+2}$	
-0.76	$2CNS^-$	Cr				-3e			$Cr(CNS)_2^+$	
-0.44	CNS^-	Cd				-2e			$CdCNS^+$	
-0.43	$4CNS^-$	Cd				-2e			$Cd(CNS)_4^{-2}$	
-0.42	$2CNS^-$	Cd				-2e			$Cd(CNS)_2$	
-0.42	$3CNS^-$	Cd				-2e			$Cd(CNS)_3^-$	
-0.39	CNS^-	In				-3e			$InCNS^{+2}$	
-0.39	CNS^-	Tl				-e			$TlCNS$	
-0.37	$3CNS^-$	In				-3e			$In(CNS)_3$	
-0.35	$4CNS^-$	Co				-2e			$Co(CNS)_4^{-2}$	
-0.35	$2CNS^-$	In				-3e			$In(CNS)_2^+$	
-0.30	$3CNS^-$	Co				-2e			$Co(CNS)_3^-$	
-0.28	CNS^-	Co				-2e			$CoCNS^+$	
-0.28	CNS^-	Ni				-2e			$NiCNS^+$	
-0.27	CNS^-	Cu				-e			$CuCNS$	
-0.26	$2CNS^-$	Co				-2e			$Co(CNS)_2$	
-0.26	$2CNS^-$	Ni				-2e			$Ni(CNS)_2$	
-0.26	$3CNS^-$	Ni				-2e			$Ni(CNS)_3^-$	
-0.02	$4CNS^-$	Cu				-e			$Cu(CNS)_4^{-3}$	
+0.01	CNS^-			UO_2^+		-e			UO_2CNS^+	
+0.02	$3CNS^-$			UO_2^+		-e			$UO_2(CNS)_3^-$	
+0.05	$2CNS^-$			UO_2^+		-e			$UO_2(CNS)_2$	
+0.18	CNS^-	Bi				-3e			$BiCNS^{+2}$	
+0.18	$2CNS^-$	Bi				-3e			$Bi(CNS)_2^+$	
+0.27	$4CNS^-$	Hg				-2e			$Hg(CNS)_4^{-2}$	
+0.65	$3CNS^-$	Ag				-e			$Ag(CNS)_3^{-2}$	
+0.66	$4CNS^-$	Au				-3e			$Au(CNS)_4^-$	
+0.74	$4CNS^-$	Ag				-e			$Ag(CNS)_4^{-3}$	

EMF	Anion	Elem	Compound or anion	Cation	H$_2$O	e =	Cation	Anion	Complex or compound	Compound or element
-2.97	$S_2O_3^{-2}$	Ba				-2e			BaS_2O_3	
-2.97	$S_2O_3^{-2}$	K				-e			$KS_2O_3^{-}$	
-2.95	$S_2O_3^{-2}$	Sr				-2e			SrS_2O_3	
-2.93	$S_2O_3^{-2}$	Ca				-2e			CaS_2O_3	
-2.75	$S_2O_3^{-2}$	Na				-e			$NaS_2O_3^{-}$	
-2.39	$S_2O_3^{-2}$	Mg				-2e			MgS_2O_3	
-1.11	$S_2O_3^{-2}$	Mn				-2e			MnS_2O_3	
-0.83	$S_2O_3^{-2}$	Zn				-2e			ZnS_2O_3	
-0.52	$S_2O_3^{-2}$	Cd				-2e			CdS_2O_3	
-0.50	$S_2O_3^{-2}$	Fe				-2e			FeS_2O_3	
-0.47	$2S_2O_3^{-2}$	Cd				-2e			$Cd(S_2O_3)_2^{-2}$	
-0.45	$S_2O_3^{-2}$	Tl				-2e			$TlS_2O_3^{-}$	
-0.34	$S_2O_3^{-2}$	Co				-2e			CoS_2O_3	
-0.31	$S_2O_3^{-2}$	Ni				-2e			NiS_2O_3	
-0.28	$2S_2O_3^{-2}$	Pb				-2e			$Pb(S_2O_3)_2^{-2}$	
-0.17	$3S_2O_3^{-2}$	Pb				-2e			$Pb(S_2O_3)_3^{-4}$	
-0.09	$S_2O_3^{-2}$	Cu				-e			$CuS_2O_3^{-}$	
-0.02	$2S_2O_3^{-2}$	Cu				-2e			$Cu(S_2O_3)_2^{-2}$	
+0.02	$2S_2O_3^{-2}$	Ag				-e			$Ag(S_2O_3)_2^{-3}$	
+0.03	$2S_2O_3^{-2}$	Hg				-2e			$Hg(S_2O_3)_2^{-2}$	
+0.28	$S_2O_3^{-2}$	Ag				-e			$AgS_2O_3^{-}$	
+0.40	$2S_2O_3^{-2}$	Cu				-e			$Cu(S_2O_3)_2^{-3}$	
+0.42	$3S_2O_3^{-2}$	Cu				-e			$Cu(S_2O_3)_3^{-5}$	

EMF	Anion	Elem	Compound or anion	Cation	H_2O	e =	Cation	Anion	Complex or compound	Compound or element
-0.45	CSN_2H_4	Cd				-2e			$Cd(CSN_2H_4)^{+2}$	
-0.43	$2CSN_2H_4$	Cd				-2e			$Cd(CSN_2H_4)_2^{+2}$	

EMF	Anion	Elem	Compound or anion	Cation	H_2O	e =	Cation	Anion	Complex or compound	Compound or element
-3.00	$P_3O_9^{-3}$	Ba				-2e			$BaP_3O_9^{-}$	
-2.99	$P_3O_9^{-3}$	Sr				-2e			$SrP_3O_9^{-}$	
-2.97	$P_3O_9^{-3}$	Ca				-2e			$CaP_3O_9^{-}$	
-2.78	$P_3O_9^{-3}$	Na				-e			$NaP_3O_9^{-2}$	
-2.63	$P_3O_9^{-3}$	La				-3e			LaP_3O_9	
-2.44	$P_3O_9^{-3}$	Mg				-2e			$MgP_3O_9^{-}$	
-1.16	$P_3O_9^{-3}$	Mn				-2e			$MnP_3O_9^{-}$	
-0.35	$P_3O_9^{-3}$	Ni				-2e			$NiP_3O_9^{-}$	

PART IV

CHEMICAL ELECTRODE POTENTIALS
of
ORGANIC FUNCTIONAL GROUPS

EMF	Elem	Compound	e =	Compound	Cation
-0.30	H	Cyclopentane-COO⁻	-e	Cyclopentane-COOH	
-0.30	H	Trimethylacetic Acid⁻	-e	Trimethylacetic Acid	
-0.29	H	Cyclohexane-COO⁻	-e	Cyclohexane-COOH	
-0.29	H	Cyclopropane-COO⁻	-e	Cyclopropane-COOH	
-0.29	H	Dimethylacetic Acid⁻	-e	Dimethylacetic Acid	
-0.29	H	n-Octanoic Acid⁻	-e	n-Octanoic Acid	
-0.29	H	Propionic Acid⁻	-e	Propionic Acid	
-0.29	H	n-Valeric Acid⁻	-e	n-Valeric Acid	
-0.28	H	Acetic Acid⁻	-e	Acetic Acid	
-0.28	H	(trans)Penta-2-en-1-oic Acid⁻	-e	(trans)Penta-2-en-1-oic Acid	
-0.27	H	(trans)Penta-3-en-1-oic Acid⁻	-e	(trans)Penta-3-en-1-oic Acid	
-0.26	H	(trans)Cinnamic Acid⁻	-e	(trans)Cinnamic Acid	
-0.26	H	Phenylacetic Acid⁻	-e	Phenylacetic Acid	
-0.25	H	Acrylic Acid⁻	-e	Acrylic Acid	
-0.23	H	(cis)Cinnamic Acid⁻	-e	(cis)Cinnamic Acid	
-0.23	H	Hydroxylacetic Acid⁻	-e	Hydroxylacetic Acid	
-0.22	H	Carbamoylacetic Acid⁻	-e	Carbamoylacetic Acid	
-0.22	H	Formic Acid⁻	-e	Formic Acid	
-0.22	H	Methylthioacetic Acid⁻	-e	Methylthioacetic Acid	
-0.21	H	Acetylacetic Acid⁻	-e	Acetylacetic Acid	
-0.21	H	Methoxyacetic Acid⁻	-e	Methoxyacetic Acid	
-0.20	H	Ethoxycarbonylacetic Acid⁻	-e	Ethoxycarbonylacetic Acid	
-0.19	H	Iodoacetic Acid⁻	-e	Iodoacetic Acid	
-0.17	H	Bromoacetic Acid⁻	-e	Bromoacetic Acid	
-0.17	H	Chloroacetic Acid⁻	-e	Chloroacetic Acid	
-0.15	H	(trans)But-2-yn-1-oic Acid⁻	-e	(trans)But-2-yn-1-oic Acid	
-0.15	H	Cyanoacetic Acid⁻	-e	Cyanoacetic Acid	
-0.15	H	Fluoroacetic Acid⁻	-e	Fluoroacetic Acid	
-0.15	H	Thiocyanatoacetic Acid⁻	-e	Thiocyanatoacetic Acid	
-0.14	H	Methylsulfonylacetic Acid⁻	-e	Methylsulfonylacetic Acid	
-0.11	H	(trans)Propiolic Acid⁻	-e	(trans)Propiolic Acid	

EMF	Elem	Compound	e =	Compound	Cation
-0.10	H	Nitroacetic Acid⁻	-e	Nitroacetic Acid	
-0.07	H	Dichloroacetic Acid⁻	-e	Dichloroacetic Acid	
-0.04	H	Trichloroacetic Acid⁻	-e	Trichloroacetic Acid	
-0.01	H	Trifluoroacetic Acid⁻	-e	Trifluoroacetic Acid	

EMF	Elem	Compound	e =	Compound	Cation
-0.27	H	p-Phenoxybenzoic Acid⁻	-e	p-Phenoxybenzoic Acid	
-0.26	H	p-Acetoxybenzoic Acid⁻	-e	p-Acetoxybenzoic Acid	
-0.26	H	p-Methylbenzoic Acid⁻	-e	p-Methylbenzoic Acid	
-0.26	H	p-Methoxybenzoic Acid⁻	-e	p-Methoxybenzoic Acid	
-0.26	H	Pyrrole-2-COO⁻	-e	Pyrrole-2-COOH	
-0.25	H	p-Acetamidobenzoic Acid⁻	-e	p-Acetamidobenzoic Acid	
-0.25	H	Benzoic Acid⁻	-e	Benzoic Acid	
-0.25	H	m-Methylbenzoic Acid⁻	-e	m-Methylbenzoic Acid	
-0.25	H	2-Naphthoic Acid⁻	-e	2-Naphthoic Acid	
-0.24	H	m-Acetamidobenzoic Acid⁻	-e	m-Acetamidobenzoic Acid	
-0.24	H	m-Acetoxybenzoic Acid⁻	-e	m-Acetoxybenzoic Acid	
-0.24	H	o-Acetylbenzoic Acid⁻	-e	o-Acetylbenzoic Acid	
-0.24	H	p-Chlorobenzoic Acid⁻	-e	p-Chlorobenzoic Acid	
-0.24	H	p-Fluorobenzoic Acid⁻	-e	p-Fluorobenzoic Acid	
-0.24	H	m-Methoxybenzoic Acid⁻	-e	m-Methoxybenzoic Acid	
-0.24	H	o-Methoxybenzoic Acid⁻	-e	o-Methoxybenzoic Acid	
-0.24	H	Thiophene-3-COO⁻	-e	Thiophene-3-COOH	
-0.23	H	m Acetylbenzoic Acid⁻	-e	m-Acetylbenzoic Acid	
-0.23	H	m-Bromobenzoic Acid⁻	-e	m-Bromobenzoic Acid	
-0.23	H	p-Bromobenzoic Acid⁻	-e	p-Bromobenzoic Acid	
-0.23	H	m-Chlorobenzoic Acid⁻	-e	m-Chlorobenzoic Acid	
-0.23	H	m-Fluorobenzoic Acid⁻	-e	m-Fluorobenzoic Acid	
-0.23	H	Furan-3-COO⁻	-e	Furan-3-COOH	
-0.23	H	m-Iodobenzoic Acid⁻	-e	m-Iodobenzoic Acid	
-0.23	H	p-Iodobenzoic Acid⁻	-e	p-Iodobenzoic Acid	
-0.23	H	o-Methylbenzoic Acid⁻	-e	o-Methylbenzoic Acid	
-0.23	H	m-Phenoxybenzoic Acid⁻	-e	m-Phenoxybenzoic Acid	
-0.22	H	p-Acetylbenzoic Acid⁻	-e	p-Acetylbenzoic Acid	
-0.22	H	1-Naphthoic Acid⁻	-e	1-Naphthoic Acid	
-0.21	H	o-Acetamidobenzoic Acid⁻	-e	o-Acetamidobenzoic Acid	
-0.21	H	o-Acetoxybenzoic Acid⁻	-e	o-Acetoxybenzoic Acid	

EMF	Elem	Compound	e =	Compound	Cation
-0.21	H	m-Cyanobenzoic Acid⁻	-e	m-Cyanobenzoic Acid	
-0.21	H	p-Cyanobenzoic Acid⁻	-e	p-Cyanobenzoic Acid	
-0.21	H	m-Nitrobenzoic Acid⁻	-e	m-Nitrobenzoic Acid	
-0.21	H	o-Phenoxybenzoic Acid⁻	-e	o-Phenoxybenzoic Acid	
-0.21	H	m-Sulfamylbenzoic Acid⁻	-e	m-Sulfamylbenzoic Acid	
-0.21	H	p-Sulfamylbenzoic Acid⁻	-e	p-Sulfamylbenzoic Acid	
-0.21	H	Thiophene-2-COO⁻	-e	Thiophene-2-COOH	
-0.20	H	p-Nitrobenzoic Acid⁻	-e	p-Nitrobenzoic Acid	
-0.20	H	o-Phenylbenzoic Acid⁻	-e	o-Phenylbenzoic Acid	
-0.19	H	o-Fluorobenzoic Acid⁻	-e	o-Fluorobenzoic Acid	
-0.19	H	Furan-2-COO⁻	-e	Furan-2-COOH	
-0.17	H	o-Bromobenzoic Acid⁻	-e	o-Bromobenzoic Acid	
-0.17	H	o-Chlorobenzoic Acid⁻	-e	o-Chlorobenzoic Acid	
-0.17	H	o-Iodobenzoic Acid⁻	-e	o-Iodobenzoic Acid	
-0.13	H	o-Nitrobenzoic Acid⁻	-e	o-Nitrobenzoic Acid	
-0.04	H	2,4,6-Trinitrobenzoic Acid⁻	-e	2,4,6-Trinitrobenzoic Acid	

EMF	Elem	Compound	e =	Compound	Cation
-0.92	H	Allyl Alcohol⁻	-e	Allyl Alcohol	
-0.92	H	Methanol⁻	-e	Methanol	
-0.89	H	Glycol⁻	-e	Glycol	
-0.88	H	Glycol methyl ether⁻	-e	Glycol methyl ether	
-0.86	H	Acetaldehyde⁻	-e	Acetaldehyde	
-0.81	H	Formaldehyde⁻	-e	Formaldehyde	
-0.80	H	Propargyl alcohol⁻	-e	Propargyl alcohol	
-0.80	H	Mannitol⁻	-e	Mannitol	
-0.75	H	Sucrose⁻	-e	Sucrose	
-0.73	H	Trifluoroethanol⁻	-e	Trifluoroethanol	
-0.72	H	Pyridine-4-aldehyde⁻	-e	Pyridine-4-aldehyde	
-0.72	H	Glucose⁻	-e	Glucose	
-0.72	H	Trichloroethanol⁻	-e	Trichloroethanol	
-0.67	H	Chloral⁻	-e	Chloral	
-0.63	H	Ethylacetoacetate⁻	-e	Ethylacetoacetate	
-0.49	H	Acetylacetone⁻	-e	Acetylacetone	
-0.49	H	Benzoylacetone⁻	-e	Benzoylacetone	
-0.44	H	Diacetylacetone⁻	-e	Diacetylacetone	
-0.34	H	Glutaconic dialdehyde⁻	-e	Glutaconic dialdehyde	
-0.34	H	Triacetylmethane⁻	-e	Triacetylmethane	
-0.31	H	Dihydroresorcinol⁻	-e	Dihydroresorcinol	

EMF	Elem	Compound	e =	Compound	Cation
-0.61	H	o-Cresol⁻	-e	o-Cresol	
-0.60	H	m-Cresol⁻	-e	m-Cresol	
-0.60	H	p-Cresol⁻	-e	p-Cresol	
-0.60	H	p-Methoxyphenol⁻	-e	p-Methoxyphenol	
-0.59	H	p-Fluorophenol⁻	-e	p-Fluorophenol	
-0.59	H	o-Methoxyphenol⁻	-e	o-Methoxyphenol	
-0.59	H	Phenol⁻	-e	Phenol	
-0.59	H	o-Phenylphenol⁻	-e	o-Phenylphenol	
-0.58	H	1-Naphthol⁻	-e	1-Naphthol	
-0.57	H	m-Methoxyphenol⁻	-e	m-Methoxyphenol	
-0.57	H	2-Naphthol⁻	-e	2-Naphthol	
-0.57	H	m-Phenylphenol⁻	-e	m-Phenylphenol	
-0.56	H	p-Chlorophenol⁻	-e	p-Chlorophenol	
-0.56	H	m-Methylthiophenol⁻	-e	m-Methylthiophenol	
-0.56	H	p-Methylthiophenol⁻	-e	p-Methylthiophenol	
-0.55	H	p-Bromophenol⁻	-e	p-Bromophenol	
-0.55	H	p-Iodophenol⁻	-e	p-Iodophenol	
-0.55	H	m-Fluorophenol⁻	-e	m-Fluorophenol	
-0.54	H	m-Acetylphenol⁻	-e	m-Acetylphenol	
-0.54	H	m-Iodophenol⁻	-e	m-Iodophenol	
-0.53	H	m-Bromophenol⁻	-e	m-Bromophenol	
-0.53	H	m-Chlorophenol⁻	-e	m-Chlorophenol	
-0.53	H	m-Formylphenol⁻	-e	m-Formylphenol	
-0.52	H	o-Fluorophenol⁻	-e	o-Fluorophenol	
-0.51	H	m-Cyanophenol⁻	-e	m-Cyanophenol	
-0.50	H	o-Chlorophenol⁻	-e	o-Chlorophenol	
-0.50	H	o-Bromophenol⁻	-e	o-Bromophenol	
-0.50	H	o-Formylphenol⁻	-e	o-Formylphenol	
-0.50	H	o-Iodophenol⁻	-e	o-Iodophenol	
-0.50	H	p-Methoxycarbonylphenol⁻	-e	p-Methoxycarbonylphenol	

EMF	Elem	Compound	e =	Compound	Cation
-0.50	H	m-Methylsulfonylphenol⁻	-e	m-Methylsulfonylphenol	
-0.50	H	m-Nitrophenol⁻	-e	m-Nitrophenol	
-0.48	H	p-Acetylphenol⁻	-e	p-Acetylphenol	
-0.47	H	p-Cyanophenol⁻	-e	p-Cyanophenol	
-0.46	H	p-Methylsulfonylphenol⁻	-e	p-Methylsulfonylphenol	
-0.45	H	p-Formylphenol⁻	-e	p-Formylphenol	
-0.43	H	o-Nitrophenol⁻	-e	o-Nitrophenol	
-0.42	H	p-Nitrophenol⁻	-e	p-Nitrophenol	
-0.04	H	Trinitrophenol⁻	-e	Trinitrophenol	

EMF	Elem	Compound	e =	Compound	Cation
-0.89	H	Acetamide⁻	-e	Acetamide	
-0.77	H	Benzamide⁻	-e	Benzamide	
-0.73	H	Acetoxime⁻	-e	Acetoxime	
-0.68	H	Glutarimide⁻	-e	Glutarimide	
-0.60	H	Nitromethane⁻	-e	Nitromethane	
-0.59	H	(b)Phenylethylboric Acid⁻	-e	(b)Phenylethylboric Acid	
-0.57	H	Succinimide⁻	-e	Succinimide	
-0.56	H	Acethydroxamic Acid⁻	-e	Acethydroxamic Acid	
-0.53	H	Benzhydroxamic Acid⁻	-e	Benzhydroxamic Acid	
-0.52	H	Phenylboric Acid⁻	-e	Phenylboric Acid	
-0.50	H	Nitroethane⁻	-e	Nitroethane	
-0.24	H	N-Cyanoacetamide⁻	-e	N-Cyanoacetamide	
-0.22	H	Cyanic Acid⁻	-e	Cyanic Acid	
-0.10	H	p-Toluenesulfinic Acid⁻	-e	p-Toluenesulfinic Acid	
-0.09	H	Benzenesulfinic Acid⁻	-e	Benzenesulfinic Acid	

EMF	Elem	Compound	e =	Compound	Cation
-0.59	H	Quinol$^-$	-e	Quinol	
-0.56	H	Catechol$^-$	-e	Catechol	
-0.56	H	Resorcinol$^-$	-e	Resorcinol	
-0.41	2H	Salicylic Acid^{-2}	-2e	Salicylic Acid	
-0.28	2H	p-Hydroxybenzoic Acid^{-2}	-2e	p-Hydroxybenzoic Acid	
-0.27	H	p-Hydroxybenzoic Acid$^-$	-e	p-Hydroxybenzoic Acid	
-0.27	2H	m-Sulfonphenol^{-2}	-2e	m-Sulfonphenol	
-0.26	H	Adipic Acid$^-$	-e	Adipic Acid	
-0.26	H	Glutaric Acid$^-$	-e	Glutaric Acid	
-0.26	2H	p-Sulfonphenol^{-2}	-2e	p-Sulfonphenol	
-0.25	H	Succinic Acid$^-$	-e	Succinic Acid	
-0.24	H	m-Hydroxybenzoic Acid$^-$	-e	m-Hydroxybenzoic Acid	
-0.21	H	Isophthalic Acid$^-$	-e	Isophthalic Acid	
-0.21	H	Terephthalic Acid$^-$	-e	Terephthalic Acid	
-0.21	2H	Phenylphosphonic Acid^{-2}	-2e	Phenylphosphonic Acid	
-0.19	H	Citric Acid$^-$	-e	Citric Acid	
-0.19	2H	Methyl phosphate^{-2}	-2e	Methyl phosphate	
-0.18	H	Fumaric Acid$^-$	-e	Fumaric Acid	
-0.18	2H	Maleic Acid^{-2}	-2e	Maleic Acid	
-0.18	H	Salicylic Acid$^-$	-e	Salicylic Acid	
-0.18	H	Tartaric Acid$^-$	-e	Tartaric Acid	
-0.17	H	Malonic Acid$^-$	-e	Malonic Acid	
-0.17	2H	Malonic Acid^{-2}	-2e	Malonic Acid	
-0.17	H	o-Phthalic Acid$^-$	-e	o-Phthalic Acid	
-0.17	2H	Succinic Acid^{-2}	-2e	Succinic Acid	
-0.16	2H	Adipic Acid^{-2}	-2e	Adipic Acid	
-0.16	2H	Glutaric Acid^{-2}	-2e	Glutaric Acid	
-0.16	2H	o-Phthalic Acid^{-2}	-2e	o-Phthalic Acid	
-0.15	H	Oxaloacetic Acid$^-$	-e	Oxaloacetic Acid	
-0.14	2H	Citric Acid^{-2}	-2e	Citric Acid	
-0.14	2H	Isophthalic Acid^{-2}	-2e	Isophthalic Acid	

EMF	Elem	Compound	e =	Compound	Cation
-0.13	3H	Citric Acid^{-3}	-3e	Citric Acid	
-0.13	2H	Fumaric Acid^{-2}	-2e	Fumaric Acid	
-0.13	2H	Oxalic Acid^{-2}	-2e	Oxalic Acid	
-0.13	2H	Oxaloacetic Acid^{-2}	-2e	Oxaloacetic Acid	
-0.13	2H	Tartaric Acid^{-2}	-2e	Tartaric Acid	
-0.13	2H	Terephthalic Acid^{-2}	-2e	Terephthalic Acid	
-0.11	H	Maleic Acid^{-}	-e	Maleic Acid	
-0.11	H	Phenylphosphonic Acid^{-}	-e	Phenylphosphonic Acid	
-0.11	2H	m-Sulfonbenzoic Acid^{-2}	-2e	m-Sulfonbenzoic Acid	
-0.11	2H	p-Sulfonbenzoic Acid^{-2}	-2e	p-Sulfonbenzoic Acid	
-0.09	H	Methyl phosphate^{-}	-e	Methyl phosphate	
-0.08	H	Mellitic Acid^{-}	-e	Mellitic Acid	
-0.08	H	Oxalic Acid^{-}	-e	Oxalic Acid	
-0.07	3H	Mellitic Acid^{-3}	-3e	Mellitic Acid	
-0.07	4H	Mellitic Acid^{-4}	-4e	Mellitic Acid	
-0.07	5H	Mellitic Acid^{-5}	-5e	Mellitic Acid	
-0.07	6H	Mellitic Acid^{-6}	-6e	Mellitic Acid	
-0.06	2H	Mellitic Acid^{-2}	-2e	Mellitic Acid	

EMF	Elem	Compound	e =	Compound	Cation
-0.02	H	2,6-Dimethyl-4-pyrone⁻	-e	2,6-Dimethyl-4-pyrone	
+0.17	H	Dioxan⁻	-e	Dioxan	
+0.21	H	Diethyl ether⁻	-e	Diethyl ether	
+0.40	H	Cyclohexanone⁻	-e	Cyclohexanone	
+0.43	H	Acetone⁻	-e	Acetone	

EMF	Elem	Compound	e =	Compound	Cation
-0.62	H	Ethyl mercaptan⁻	-e	Ethyl mercaptan	
-0.62	H	Thioglycollic Acid (S⁻)	-e	Thioglycollic Acid	
-0.60	H	2-Mercaptopropionic Acid (S⁻)	-e	2-Mercaptopropionic Acid	
-0.56	H	Benzylmercaptan⁻	-e	Benzylmercaptan	
-0.56	H	(b)Hydroxyethylmercaptan⁻	-e	(b)Hydroxyethylmercaptan	
-0.56	H	2-Mercaptoethane sulfonic Acid⁻	-e	2-Mercaptoethane sulfonic Acid	
-0.45	H	Methyl thioglycollate⁻	-e	Methyl thioglycollate	
-0.38	H	Phenylmercaptan⁻	-e	Phenylmercaptan	
-0.26	H	2-Mercaptopropionic Acid⁻	-e	2-Mercaptopropionic Acid	
-0.22	H	Thioglycollic Acid⁻	-e	Thioglycollic Acid	
+0.23		$(CH_3)_2SO$ + H_2O	-2e	$(CH_3)_2SO_2$	$2H^+$

EMF	Elem	Compound	e =	Compound	Cation
-0.66	H	Malonitrile⁻	-e	Malonitrile	
-0.61	H	Cyanamide⁻	-e	Cyanamide	
-0.32	H	Dipicrylamine⁻	-e	Dipicrylamine	
-0.06	H	Dicyanamide⁻	-e	Dicyanamide	

EMF	Elem	Compound	e =	Compound	Cation
-0.80	H	N,N'-Dimethylguanidine⁻	-e	N,N'-Dimethylguanidine	
-0.80	H	Guanidine⁻	-e	Guanidine	
-0.79	H	N,N-Dimethylguanidine⁻	-e	N,N-Dimethylguanidine	
-0.79	H	N-Methylguanidine⁻	-e	N-Methylguanidine	
-0.76	H	Diguanide⁻	-e	Diguanide	
-0.73	H	Acetamidine⁻	-e	Acetamidine	
-0.69	H	Benzamidine⁻	-e	Benzamidine	
-0.67	H	Azetidine⁻	-e	Azetidine	
-0.67	H	Pyrrolidine⁻	-e	Pyrrolidine	
-0.66	H	Piperidine⁻	-e	Piperidine	
-0.65	H	Diethylamine⁻	-e	Diethylamine	
-0.64	H	Dimethylamine⁻	-e	Dimethylamine	
-0.64	H	Triethylamine⁻	-e	Triethylamine	
-0.63	H	n-Butylamine⁻	-e	n-Butylamine	
-0.63	H	Cyclohexylamine⁻	-e	Cyclohexylamine	
-0.63	H	Docosylamine⁻	-e	Docosylamine	
-0.63	H	Dodecylamine⁻	-e	Dodecylamine	
-0.63	H	Ethylamine⁻	-e	Ethylamine	
-0.63	H	Hexadecylamine⁻	-e	Hexadecylamine	
-0.63	H	Methylamine⁻	-e	Methylamine	
-0.63	H	Octylamine⁻	-e	Octylamine	
-0.63	H	iso-Propylamine⁻	-e	iso-Propylamine	
-0.63	H	Undecylamine⁻	-e	Undecylamine	
-0.62	H	tert-Butylamine⁻	-e	tert-Butylamine	
-0.62	H	N-Methylpyrrolidine⁻	-e	N-Methylpyrrolidine	
-0.62	H	n-Propylamine⁻	-e	n-Propylamine	
-0.60	H	N-Methylpiperidine⁻	-e	N-Methylpiperidine	
-0.58	H	O-Methyl-iso-urea⁻	-e	O-Methyl-iso-urea	
-0.58	H	S-Methyl-iso-thiourea⁻	-e	S-Methyl-iso-thiourea	
-0.58	H	Phenylethylamine⁻	-e	Phenylethylamine	
-0.58	H	Trimethylamine⁻	-e	Trimethylamine	

EMF	Elem	Compound	e =	Compound	Cation
-0.57	H	Allylmethylamine$^-$	-e	Allylmethylamine	
-0.57	H	2-Methyl-(D)^2-tetrahydro-pyridine$^-$	-e	2-Methyl-(D)^2-tetrahydro-pyridine	
-0.56	H	Ethanolamine$^-$	-e	Ethanolamine	
-0.56	H	Methoxyethylamine$^-$	-e	Methoxyethylamine	
-0.55	H	Benzylmethylamine$^-$	-e	Benzylmethylamine	
-0.54	H	Cyclohexanonimine$^-$	-e	Cyclohexanonimine	
-0.54	H	Ethoxycarbonylethylamine$^-$	-e	Ethoxycarbonylethylamine	
-0.51	H	Morpholine$^-$	-e	Morpholine	
-0.49	H	N-Acetylguanidine$^-$	-e	N-Acetylguanidine	
-0.48	H	Aziridine$^-$	-e	Aziridine	
-0.47	H	Carbamylmethylamine$^-$	-e	Carbamylmethylamine	
-0.46	H	Cyanoethylamine$^-$	-e	Cyanoethylamine	
-0.46	H	Triethanolamine$^-$	-e	Triethanolamine	
-0.45	H	Methoxycarbonylmethylamine$^-$	-e	Methoxycarbonylmethylamine	
-0.40	H	Diphenyl Ketimine$^-$	-e	Diphenyl Ketimine	
-0.32	H	Cyanomethylamine$^-$	-e	Cyanomethylamine	
-0.19	H	Acethydrazide$^-$	-e	Acethydrazide	
-0.07	H	Cyanamide$^-$	-e	Cyanamide	
-0.01	H	Urea$^-$	-e	Urea	
+0.01	2H	Acetamide^{-2}	-2e	Acetamide	
+0.03	H	Acetamide$^-$	-e	Acetamide	
+0.06	H	Thiourea$^-$	-e	Thiourea	

EMF	Elem	Compound	e =	Compound	Cation
-0.60	H	sym-Diphenylguanidine⁻	-e	sym-Diphenylguanidine	
-0.54	H	4-Aminopyridine⁻	-e	4-Aminopyridine	
-0.42	H	N-Diethyl-o-toluidine⁻	-e	N-Diethyl-o-toluidine	
-0.42	H	Imidazole⁻	-e	Imidazole	
-0.42	H	N-tertButylaniline⁻	-e	N-tertButylaniline	
-0.41	H	2-Aminopyridine⁻	-e	2-Aminopyridine	
-0.39	H	N-Diethylaniline	-e	N-Diethylaniline	
-0.39	H	4-Methoxypyridine⁻	-e	4-Methoxypyridine	
-0.36	H	4-Methylpyridine⁻	-e	4-Methylpyridine	
-0.35	H	3-Aminopyridine⁻	-e	3-Aminopyridine	
-0.35	H	N-Dimethyl-o-toluidine⁻	-e	N-Dimethyl-o-toluidine	
-0.35	H	2-Methylpyridine⁻	-e	2-Methylpyridine	
-0.34	H	3-Methylpyridine⁻	-e	3-Methylpyridine	
-0.34	H	2-tert-Butylpyridine⁻	-e	2-tert-Butylpyridine	
-0.33	H	Acridine⁻	-e	Acridine	
-0.33	H	Benzimidazole⁻	-e	Benzimidazole	
-0.33	H	N-isopropylaniline⁻	-e	N-isopropylaniline	
-0.31	H	p-Ethoxyaniline⁻	-e	p-Ethoxyaniline	
-0.31	H	p-Methoxyaniline⁻	-e	p-Methoxyaniline	
-0.31	H	Pyridine⁻	-e	Pyridine	
-0.30	H	N-Dimethylaniline⁻	-e	N-Dimethylaniline	
-0.30	H	N-Ethylaniline⁻	-e	N-Ethylaniline	
-0.30	H	p-Methylaniline⁻	-e	p-Methylaniline	
-0.29	H	3-Methoxypyridine⁻	-e	3-Methoxypyridine	
-0.29	H	N-Methylaniline⁻	-e	N-Methylaniline	
-0.29	H	Quinoline⁻	-e	Quinoline	
-0.28	H	p-Fluoroaniline⁻	-e	p-Fluoroaniline	
-0.28	H	m-Methylaniline⁻	-e	m-Methylaniline	
-0.27	H	Aniline⁻	-e	Aniline	
-0.27	H	o-Methoxyaniline⁻	-e	o-Methoxyaniline	
-0.26	H	o-Ethoxyaniline⁻	-e	o-Ethoxyaniline	

EMF	Elem	Compound	e =	Compound	Cation
-0.26	H	o-Methylaniline⁻	-e	o-Methylaniline	
-0.26	H	p-Methylthioaniline⁻	-e	p-Methylthioaniline	
-0.25	H	m-Ethoxyaniline⁻	-e	m-Ethoxyaniline	
-0.25	H	m-Methoxyaniline⁻	-e	m-Methoxyaniline	
-0.25	H	m-Phenylaniline⁻	-e	m-Phenylaniline	
-0.25	H	p-Phenylaniline⁻	-e	p-Phenylaniline	
-0.24	H	p-Chloroaniline⁻	-e	p-Chloroaniline	
-0.24	H	m-Methylthioaniline⁻	-e	m-Methylthioaniline	
-0.24	H	2-Naphthylamine⁻	-e	2-Naphthylamine	
-0.23	H	p-Bromoaniline⁻	-e	p-Bromoaniline	
-0.23	H	2,6-Dimethylaniline⁻	-e	2,6-Dimethylaniline	
-0.23	H	1-Naphthylamine⁻	-e	1-Naphthylamine	
-0.22	H	p-Iodoaniline⁻	-e	p-Iodoaniline	
-0.22	H	m-Methoxycarbonylaniline⁻	-e	m-Methoxycarbonylaniline	
-0.22	H	o-Phenylaniline⁻	-e	o-Phenylaniline	
-0.22	H	2-tert-Butylaniline⁻	-e	2-tert-Butlyaniline	
-0.21	H	m-Bromoaniline⁻	-e	m-Bromoaniline	
-0.21	H	m-Fluoroaniline⁻	-e	m-Fluoroaniline	
-0.21	H	m-Iodoaniline⁻	-e	m-Iodoaniline	
-0.21	H	Phthalazine⁻	-e	Phthalazine	
-0.21	H	m-Trifluoromethylaniline⁻	-e	m-Trifluoromethylaniline	
-0.20	H	m-Chloroaniline⁻	-e	m-Chloroaniline	
-0.19	H	3-Acetylpyridine⁻	-e	3-Acetylpyridine	
-0.19	H	o-Fluoroaniline⁻	-e	o-Fluoroaniline	
-0.19	H	2-Methoxypyridine⁻	-e	2-Methoxypyridine	
-0.17	H	3-Chloropyridine⁻	-e	3-Chloropyridine	
-0.16	H	o-Chloroaniline⁻	-e	o-Chloroaniline	
-0.16	H	m-Cyanoaniline⁻	-e	m-Cyanoaniline	
-0.16	H	m-Methylsulfonylaniline⁻	-e	m-Methylsulfonylaniline	
-0.15	H	o-Bromoaniline⁻	-e	o-Bromoaniline	
-0.15	H	o-Iodoaniline⁻	-e	o-Iodoaniline	

EMF	Elem	Compound	e =	Compound	Cation
-0.15	H	m-Nitroaniline⁻	-e	m-Nitroaniline	
-0.15	H	Pyrazole⁻	-e	Pyrazole	
-0.15	H	Thiazole⁻	-e	Thiazole	
-0.15	H	p-Trifluoromethylaniline⁻	-e	p-Trifluoromethylaniline	
-0.14	H	p-Methoxycarbonylaniline⁻	-e	p-Methoxycarbonylaniline	
-0.14	H	Pyridazine⁻	-e	Pyridazine	
-0.13	H	o-Methoxycarbonylaniline⁻	-e	o-Methoxycarbonylaniline	
-0.11	H	2-Iodopyridine⁻	-e	2-Iodopyridine	
-0.10	H	p-Cyanoaniline⁻	-e	p-Cyanoaniline	
-0.09	H	Benztriazole⁻	-e	Benztriazole	
-0.09	H	3-Cyanopyridine⁻	-e	3-Cyanopyridine	
-0.09	H	p-Methylsulfonylaniline⁻	-e	p-Methylsulfonylaniline	
-0.08	H	Pyrimidine⁻	-e	Pyrimidine	
-0.07	H	Phenazine⁻	-e	Phenazine	
-0.06	H	p-Nitroaniline⁻	-e	p-Nitroaniline	
-0.05	H	2-Bromopyridine⁻	-e	2-Bromopyridine	
-0.05	H	Diphenylamine⁻	-e	Diphenylamine	
-0.05	H	Pyridine-N-Oxide⁻	-e	Pyridine-N-Oxide	
-0.04	H	2-Chloropyridine⁻	-e	2-Chloropyridine	
-0.04	H	Pyrazine⁻	-e	Pyrazine	
-0.04	H	Quinoxaline⁻	-e	Quinoxaline	
-0.02	H	Acetanilide⁻	-e	Acetanilide	
+0.02	H	o-Nitroaniline⁻	-e	o-Nitroaniline	
+0.02	H	Pyrrole⁻	-e	Pyrrole	
+0.03	H	2-Fluoropyridine⁻	-e	2-Fluoropyridine	

EMF	Elem	Compound	e =	Compound	Cation
-0.65	H	1,8-Octanediamine$^-$	-e	1,8-Octanediamine	
-0.64	H	1,4-Butanediamine$^-$	-e	1,4-Butanediamine	
-0.63	H	1,3-Propanediamine$^-$	-e	1,3-Propanediamine	
-0.60	H	1,2-Ethanediamine$^-$	-e	1,2-Ethanediamine	
-0.59	H	Cis(trans)-1,2-cyclohexane-diamine$^-$	-e	Cis(trans)-1,2-cyclohexane-diamine	
-0.58	H	Piperazine$^-$	-e	Piperazine	
-0.57	H	1,3-Diamino-2-propanol$^-$	-e	1,3-Diamino-2-propanol	
-0.38	H	N,N'-Tetramethyl-p-phenylene-diamine$^-$	-e	N,N'-Tetramethyl-p-phenylene-diamine	
-0.36	H	p-Phenylenediamine$^-$	-e	p-Phenylenediamine	
-0.30	2H	1,8-Octanediamine^{-2}	-2e	1,8-Octanediamine	
-0.29	H	m-Phenylenediamine$^-$	-e	m-Phenylenediamine	
-0.28	H	Benzidine$^-$	-e	Benzidine	
-0.28	2H	1,4-Butanediamine^{-2}	-2e	1,4-Butanediamine	
-0.26	H	o-Phenylenediamine$^-$	-e	o-Phenylenediamine	
-0.26	2H	1,3-Propanediamine^{-2}	-2e	1,3-Propanediamine	
-0.23	2H	1,3-Diamino-2-propanol^{-2}	-2e	1,3-Diamino-2-propanol	
-0.21	2H	1,2-Ethanediamine^{-2}	-2e	1,2-Ethanediamine	
-0.19	2H	Cis(trans)-1,2-cyclohexane-diamine^{-2}	-2e	Cis(trans)-1,2-cyclohexane-diamine	
-0.17	2H	Piperazine^{-2}	-2e	Piperazine	
-0.11	2H	Benzidine^{-2}	-2e	Benzidine	
-0.10	2H	p-Phenylenediamine^{-2}	-2e	p-Phenylenediamine	
-0.08	2H	m-Phenylenediamine^{-2}	-2e	m-Phenylenediamine	
-0.07	2H	N,N'-Tetramethyl-p-phenylene-diamine^{-2}	-2e	N,N'-Tetramethyl-p-phenylene-diamine	
-0.06	2H	o-Phenylenediamine^{-2}	-2e	o-Phenylenediamine	

EMF	Elem	Compound	e =	Compound	Cation
-0.69		2-Hydroxyquinoline	-e	2-Hydroxyquinoline⁻	H⁺
-0.66		4-Hydroxyquinoline	-e	4-Hydroxyquinoline⁻	H⁺
-0.64		5-Aminovaleric Acid	-e	5-Aminovaleric Acid⁻	H⁺
-0.62		2-Mercaptoethylamine	-e	2-Mercaptoethylamine⁻	H⁺
-0.61		4-Aminophenol	-e	4-Aminophenol⁻	H⁺
-0.61		3-Aminopropionic Acid	-e	3-Aminopropionic Acid⁻	H⁺
-0.60		2-Mercaptoquinoline	-e	2-Mercaptoquinoline⁻	H⁺
-0.59		2-Aminopropionic Acid	-e	2-Aminopropionic Acid⁻	H⁺
-0.59		8-Hydroxyquinoline	-e	8-Hydroxyquinoline⁻	H⁺
-0.58		2-Aminoacetic Acid	-e	2-Aminoacetic Acid⁻	H⁺
-0.58		3-Aminophenol	-e	3-Aminophenol⁻	H⁺
-0.57		2-Aminophenol	-e	2-Aminophenol⁻	H⁺
-0.54		Aminoethane-(b)-sulfonic Acid	-e	Aminoethane-(b)-SO_3H^-	H⁺
-0.53		6-Hydroxyquinoline	-e	6-Hydroxyquinoline⁻	H⁺
-0.52		4-Mercaptoquinoline	-e	4-Mercaptoquinoline⁻	H⁺
-0.52		7-Hydroxyquinoline	-e	7-Hydroxyquinoline⁻	H⁺
-0.51		5-Hydroxyquinoline	-e	5-Hydroxyquinoline⁻	H⁺
-0.49	H	2-Mercaptoethylamine⁻	-e	2-Mercaptoethylamine	
-0.48		3-Hydroxyquinoline	-e	3-Hydroxyquinoline⁻	H⁺
-0.36		3-Mercaptoquinoline	-e	3-Mercaptoquinoline⁻	H⁺
-0.34		Aminomethanesulfonic Acid	-e	Aminomethanesulfonic Acid⁻	H⁺
-0.33	H	4-Aminophenol⁻	-e	4-Aminophenol	
-0.32	H	7-Hydroxyquinoline⁻	-e	7-Hydroxyquinoline	
-0.32		Pyridine-2-COOH	-e	Pyridine-2-COO⁻	H⁺
-0.31	H	5-Hydroxyquinoline⁻	-e	5-Hydroxyquinoline	
-0.31	H	6-Hydroxyquinoline⁻	-e	6-Hydroxyquinoline	
-0.30	H	8-Hydroxyquinoline⁻	-e	8-Hydroxyquinoline	
-0.29		2-Aminobenzoic Acid	-e	2-Aminobenzoic Acid⁻	H⁺
-0.29		4-Aminobenzoic Acid	-e	4-Aminobenzoic Acid⁻	H⁺
-0.29		Pyridine-4-COOH	-e	Pyridine-4-COO⁻	H⁺
-0.28		3-Aminobenzoic Acid	-e	3-Aminobenzoic Acid⁻	H⁺

EMF	Elem	Compound	e =	Compound	Cation
-0.28	H	2-Aminophenol$^-$	-e	2-Aminophenol	
-0.28		Pyridine-3-COOH	-e	Pyridine-3-COO$^-$	H$^+$
-0.25	H	3-Aminophenol$^-$	-e	3-Aminophenol	
-0.25	H	5-Aminovaleric Acid$^-$	-e	5-Aminovaleric Acid	
-0.25	H	3-Hydroxyquinoline$^-$	-e	3-Hydroxyquinoline	
-0.22		Aniline-3-SO$_3$H	-e	Aniline-3-SO$_3^-$	H$^+$
-0.21	H	3-Aminopropionic Acid$^-$	-e	3-Aminopropionic Acid	
-0.18	H	3-Aminobenzoic Acid$^-$	-e	3-Aminobenzoic Acid	
-0.18		Aniline-4-SO$_3$H	-e	Aniline-4-SO$_3^-$	H$^+$
-0.14	H	4-Aminobenzoic Acid$^-$	-e	4-Aminobenzoic Acid	
-0.14	H	3-Mercaptoquinoline$^-$	-e	3-Mercaptoquinoline	
-0.13	H	2-Aminoacetic Acid$^-$	-e	2-Aminoacetic Acid	
-0.13	H	2-Aminopropionic Acid$^-$	-e	2-Aminopropionic Acid	
-0.13	H	4-Hydroxyquinoline$^-$	-e	4-Hydroxyquinoline	
-0.12	H	2-Aminobenzoic Acid$^-$	-e	2-Aminobenzoic Acid	
-0.12	H	Pyridine-3-COO$^-$	-e	Pyridine-3-COOH	
-0.10	H	Pyridine-4-COO$^-$	-e	Pyridine-4-COOH	
-0.09	H	Aminoethane-(b)-SO$_3^-$	-e	Aminoethane-(b)-SO$_3$H	
-0.06	H	Pyridine-2-COO$^-$	-e	Pyridine-2-COOH	
-0.05	H	4-Mercaptoquinoline$^-$	-e	4-Mercaptoquinoline	
-0.03	H	Aniline-4-SO$_3^-$	-e	Aniline-4-SO$_3$H	
-0.03	H	Aniline-3-SO$_3^-$	-e	Aniline-3-SO$_3$H	
+0.02	H	2-Hydroxyquinoline$^-$	-e	2-Hydroxyquinoline	
+0.09	H	2-Mercaptoquinoline$^-$	-e	2-Mercaptoquinoline	

EMF	Elem	Compound	e =	Compound	Cation
-0.27	H	Phenol-m-sulfonate-indo-2,6-dibromophenol⁻	-e	Phenol-m-sulfonate-indo-2,6-dibromophenol	
-0.25	H	m-Bromophenol-indophenol⁻	-e	m-Bromophenol-indophenol	
-0.25	H	m-Chlorophenol-indo-2,6-dichlorophenol⁻	-e	m-Chlorophenol-indo-2,6-dichlorophenol	
-0.23	H	o-Chlorophenol-indophenol⁻	-e	o-Chlorophenol-indophenol	
-0.22	H	Bindshedler's green⁻	-e	Bindshedler's green	
-0.22	H	2,6-Dichlorophenol-indophenol⁻	-e	2,6-Dichlorophenol-indophenol	
-0.18	H	2,6-Dichlorophenol-indo-o-cresol⁻	-e	2,6-Dichlorophenol-indo-o-cresol	
-0.12	H	1-Naphthol-2-sulfonate-indophenol⁻	-e	1-Naphthol-2-sulfonate-indophenol	
-0.12	H	1-Naphthol-2-sulfonate-indo-2,6-dichlorophenol⁻	-e	1-Naphthol-2-sulfonate-indo-2,6-dichlorophenol	
-0.12	H	Toluylene Blue⁻	-e	Toluylene Blue	
-0.05	H	Cresyl Blue⁻	-e	Cresyl Blue	
-0.02	H	Gallocyanine⁻	-e	Gallocyanine	
-0.01	H	Methylene Blue⁻	-e	Methylene Blue	
+0.04	H	Ciba Scarlet Sulfonate⁻	-e	Ciba Scarlet Sulfonate	
+0.05	H	Indigo tetrasulfonate⁻	-e	Indigo tetrasulfonate	
+0.06	H	Methyl Capri Blue⁻	-e	Methyl Capri Blue	
+0.08	H	Indigo Trisulfonate⁻	-e	Indigo Trisulfonate	
+0.13	H	Indigo Disulfonate⁻	-e	Indigo Disulfonate	
+0.14	H	Gallophenine⁻	-e	Gallophenine	
+0.17	H	Brilliant Alizarine Blue⁻	-e	Brilliant Alizarine Blue	
+0.25	H	Phenosafranine⁻	-e	Phenosafranine	
+0.27	H	Tetramethylphenosafranine⁻	-e	Tetramethylphenosafranine	
+0.29	H	Safranine T⁻	-e	Safranine T	
+0.30	H	Induline Scarlet⁻	-e	Induline Scarlet	
+0.33	H	Neutral Red⁻	-e	Neutral Red	
+0.39	H	Rosindone Sulfonate No. 6⁻	-e	Rosidone Sulfonate No. 6	

EMF	Elem	Compound	e =	Compound	Cation
-0.97	H	Dicyano-1,4-benzoquinone⁻	-e	Dicyano-1,4-benzoquinone	
-0.83	H	Tetrachloro-1,2-benzoquinone⁻	-e	Tetrachloro-1,2-benzoquinone	
-0.82	H	Tetrabromo-1,2-benzoquinone⁻	-e	Tetrabromo-1,2-benzoquinone	
-0.78	H	1,2-Benzoquinone⁻	-e	1,2-Benzoquinone	
-0.72	H	2,5-Dibromo-1,4-benzoquinone⁻	-e	2,5-Dibromo-1,4-benzoquinone	
-0.72	H	2,5-Dichloro-1,4-benzoquinone⁻	-e	2,5-Dichloro-1,4-benzoquinone	
-0.72	H	2,6-Dichloro-1,4-benzoquinone⁻	-e	2,6-Dichloro-1,4-benzoquinone	
-0.71	H	Monochloro-1,4-benzoquinone⁻	-e	Monochloro-1,4-benzoquinone	
-0.71	H	Monobromo-1,4-benzoquinone⁻	-e	Monobromo-1,4-benzoquinone	
-0.71	H	2,3-Dichloro-1,4-benzoquinone⁻	-e	2,3-Dichloro-1,4-benzoquinone	
-0.66	H	2-Bromo-5-methyl-1,4-benzo-quinone⁻	-e	2-Bromo-5-methyl-1,4-benzo-quinone	
-0.65	H	2-Chloro-5-methyl-1,4-benzo-quinone⁻	-e	2-Chloro-5-methyl-1,4-benzo-quinone	
-0.64	H	2-Methyl-1,4-benzoquinone⁻	-e	2-Methyl-1,4-benzquinone	
-0.59	H	2,5-Dimethoxy-1,4-benzoquinone⁻	-e	2,5-Dimethoxy-1,4-benzoquinone	
-0.59	H	2,3-Dimethyl-1,4-benzoquinone⁻	-e	2,3-Dimethyl-1,4-benzoquinone	
-0.59	H	2,5-Dimethyl-1,4-benzoquinone⁻	-e	2,5-Dimethyl-1,4-benzoquinone	
-0.59	H	Hydroxy 1,4 benzoquinone⁻	-e	Hydroxy-1,4-benzoquinone	
-0.59	H	Monochloroxylo-1,4-benzo-quinone⁻	-e	Monochloroxylo-1,4-benzo-quinone	
-0.59	H	2-Methyl-5-isopropyl-1,4-benzoquinone⁻	-e	2-Methyl-5-isopropyl-1,4-benzoquinone	
-0.53	H	2,3,5-Trimethyl-1,4-benzo-quinone⁻	-e	2,3,5-Trimethyl-1,4-benzo-quinone	
-0.51	H	2,6-Dimethoxy-1,4-benzoquinone⁻	-e	2,6-Dimethoxy-1,4-benzoquinone	
-0.46	H	2,5-Diethoxy-1,4-benzoquinone⁻	-e	2,5-Diethoxy-1,4-benzoquinone	
-0.44	H	2,5-Dihydroxy-1,4-benzoquinone⁻	-e	2,5-Dihydroxy-1,4-benzoquinone	
-0.44	H	2,5-Dihydroxy-3,6-dichloro-1,4-benzoquinone⁻	-e	2,5-Dihydroxy-3,6-dichloro-1,4-benzoquinone	

EMF	Elem	Compound	e =	Compound	Cation
-0.24	H	9,10-Anthraquinone-1,5-disulfonic Acid⁻	-e	9,10-Anthraquinone-1,5-disulfonic Acid	
-0.23	H	9,10-Anthraquinone-2,6-disulfonic Acid⁻	-e	9,10-Anthraquinone-2,6-disulfonic Acid	
-0.23	H	9,10-Anthraquinone-2,7-disulfonic Acid⁻	-e	9,10-Anthraquinone-2,7-disulfonic Acid	
-0.21	H	9,10-Anthraquinone-1,8-disulfonic Acid⁻	-e	9,10-Anthraquinone-1,8-disulfonic Acid	
-0.20	H	9,10-Anthraquinone-1-sulfonic Acid⁻	-e	9,10-Anthraquinone-1-sulfonic Acid	
-0.19	H	9,10-Anthraquinone-2-sulfonic Acid⁻	-e	9,10-Anthraquinone-2-sulfonic Acid	

EMF	Elem	Compound	e =	Compound	Cation
-0.66	H	1,2-Naphthoquinone-4,6-disulfonate⁻	-e	1,2-Naphthoquinone-4,6-disulfonate	
-0.63	H	1,2-Naphthoquinone-4-sulfonate⁻	-e	1,2-Naphthoquinone-4-sulfonate	
-0.55	H	1,2-Naphthoquinone⁻	-e	1,2-Naphthoquinone	
-0.53	H	1,4-Naphthoquinone-2-sulfonate⁻	-e	1,4-Naphthoquinone-2-sulfonate	
-0.53	H	1,4-Naphthoquinone-3-sulfonate⁻	-e	1,4-Naphthoquinone-3-sulfonate	
-0.47	H	1,4-Naphthoquinone⁻	-e	1,4-Naphthoquinone	
-0.29	H	Dimethyl-1,4-Naphthoquinone⁻	-e	Dimethyl-1,4-Naphthoquinone	

PART V

CHEMICAL ELECTRODE POTENTIALS

OF

INORGANIC NITROGEN BASES

EMF	Base	Elem	Compound or anion	Cation	H₂O	e =	Cation	Anion	Complex or compound	Compound or element
-2.35	NH_3	Mg				-2e			$Mg(NH_3)^{+2}$	
-2.34	$2NH_3$	Mg				-2e			$Mg(NH_3)_2^{+2}$	
-2.33	$3NH_3$	Mg				-2e			$Mg(NH_3)_3^{+2}$	
-2.32	$4NH_3$	Mg				-2e			$Mg(NH_3)_4^{+2}$	
-2.31	$5NH_3$	Mg				-2e			$Mg(NH_3)_5^{+2}$	
-2.30	$6NH_3$	Mg				-2e			$Mg(NH_3)_6^{+2}$	
-1.07	NH_3	Mn				-2e			$Mn(NH_3)^{+2}$	
-1.06	$2NH_3$	Mn				-2e			$Mn(NH_3)_2^{+2}$	
-1.04	$4NH_3(aq)$	Zn				-2e			$Zn(NH_3)_4^{+2}$	
-0.83	NH_3	Zn				-2e			$Zn(NH_3)^{+2}$	
-0.83	$2NH_3$	Zn				-2e			$Zn(NH_3)_2^{+2}$	
-0.83	$3NH_3$	Zn				-2e			$Zn(NH_3)_3^{+2}$	
-0.61	$4NH_3(aq)$	Cd				-2e			$Cd(NH_3)_4^{+2}$	
-0.48	NH_3	Cd				-2e			$Cd(NH_3)^{+2}$	
-0.48	NH_3	Fe				-2e			$Fe(NH_3)^{+2}$	
-0.46	$2NH_3$	Cd				-2e			$Cd(NH_3)_2^{+2}$	
-0.46	$2NH_3$	Fe				-2e			$Fe(NH_3)_2^{+2}$	
-0.44	$3NH_3$	Cd				-2e			$Cd(NH_3)_3^{+2}$	
-0.43	$4NH_3$	Cd				-2e			$Cd(NH_3)_4^{+2}$	
-0.39	$5NH_3$	Cd				-2e			$Cd(NH_3)_5^{+2}$	
-0.35	$6NH_3$	Cd				-2e			$Cd(NH_3)_6^{+2}$	
-0.34	NH_3	Co				-2e			$Co(NH_3)^{+2}$	
-0.33	$2NH_3$	Co				-2e			$Co(NH_3)_2^{+2}$	
-0.33	NH_3	Ni				-2e			$Ni(NH_3)^{+2}$	
-0.32	$2NH_3$	Ni				-2e			$Ni(NH_3)_2^{+2}$	
-0.31	$3NH_3$	Co				-2e			$Co(NH_3)_3^{+2}$	
-0.30	$4NH_3$	Co				-2e			$Co(NH_3)_4^{+2}$	
-0.30	$3NH_3$	Ni				-2e			$Ni(NH_3)_3^{+2}$	
-0.29	$5NH_3$	Co				-2e			$Co(NH_3)_5^{+2}$	
-0.29	$4NH_3$	Ni				-2e			$Ni(NH_3)_4^{+2}$	
-0.29	NH_3	Tl				-e			$Tl(NH_3)^{+}$	
-0.27	$5NH_3$	Ni				-2e			$Ni(NH_3)_5^{+2}$	

EMF	Base	Elem	Compound or anion	Cation	H_2O	e =	Cation	Anion	Complex or compound	Compound or element
-0.26	$6NH_3$	Co				-2e			$Co(NH_3)_6^{+2}$	
-0.25	$6NH_3$	Ni				-2e			$Ni(NH_3)_6^{+2}$	
-0.22	NH_3	Cu				-2e			$Cu(NH_3)^{+2}$	
-0.12	$2NH_3$	Cu				-e			$Cu(NH_3)_2^{+}$	
+0.03	$3NH_3$	Cu				-2e			$Cu(NH_3)_3^{+2}$	
+0.15	NH_3	Cu				-e			$Cu(NH_3)^{+}$	
+0.24	$2NH_3$	Cu				-e			$Cu(NH_3)_2^{+}$	
+0.24	$2NH_3$	Cu				-2e			$Cu(NH_3)_2^{+2}$	
+0.28	$4NH_3$	Cu				-2e			$Cu(NH_3)_4^{+2}$	
+0.35	$5NH_3$	Cu				-2e			$Cu(NH_3)_5^{+2}$	
+0.57	$2NH_3$	Ag				-e			$Ag(NH_3)_2^{+}$	
+0.61	NH_3	Ag				-e			$Ag(NH_3)^{+}$	

EMF	Base	Elem	Compound or anion	Cation	H_2O	e =	Cation	Anion	Complex or compound	Compound or element
-0.83	N_2H_4	Zn				-2e			$Zn(N_2H_4)^{+2}$	
-0.81	$2N_2H_4$	Zn				-2e			$Zn(N_2H_4)_2^{+2}$	
-0.80	$3N_2H_4$	Zn				-2e			$Zn(N_2H_4)_3^{+2}$	
-0.78	$4N_2H_4$	Zn				-2e			$Zn(N_2H_4)_4^{+2}$	
-0.33	N_2H_4	Ni				-2e			$Ni(N_2H_4)^{+2}$	
-0.32	$2N_2H_4$	Ni				-2e			$Ni(N_2H_4)_2^{+2}$	
-0.31	$3N_2H_4$	Ni				-2e			$Ni(N_2H_4)_3^{+2}$	
-0.30	$4N_2H_4$	Ni				-2e			$Ni(N_2H_4)_4^{+2}$	
-0.30	$5N_2H_4$	Ni				-2e			$Ni(N_2H_4)_5^{+2}$	
-0.29	$6N_2H_4$	Ni				-2e			$Ni(N_2H_4)_6^{+2}$	

ABBREVIATIONS

Abbreviation	Name
(a)	Alpha
2AAa	2-Aminoacetic Acid
2ABA	2-Aminobenzoic Acid
3ABA	3-Aminobenzoic Acid
4ABA	4-Aminobenzoic Acid
Ac	Acetic Acid
AcAc	Acetylacetone
AcAcA	Acetylacetic Acid
ACD	Acetaldehyde
AcHz	Acethydrazide
Acm	Acetamidine
Acox	Acetoxime
3AcPy	3-Acetylpyridine
Acrd	Acridine
Acrl	Acrylic Acid
Actd	Acetamide
Actld	Acetanilide
Acto	Acetone
Adip	Adipic Acid
AESA	Aminoethane-(b)-sulfonic Acid
Af	o-Aminophenol
Ahoa	Acethydroxamic Acid
Alan	Alanine
Alma	Allylmethylamine
AlOH	Allyl alcohol
Amac	Aminobarbituric-N,N-diacetic Acid
AMSA	Aminomethanesulfonic Acid
ANIL	Aniline
2AP	2-Aminophenol
3AP	3-Aminophenol
4AP	4-Aminophenol
Ap	Malate

Abbreviation	Name
3APA	3-Aminopropionic Acid
2Apa	2-Aminopropionic Acid
2APy	2-Aminopyridine
3APy	3-Aminopyridine
4APy	4-Aminopyridine
9AQDS	9,10-Anthraquinone-1,8-disulfonic Acid
9AQ5DS	9,10-Anthraquinone-1,5-disulfonic Acid
9AQ1S	9,10-Anthraquinone-1-sulfonic Acid
9AQ2S	9,10-Anthraquinone-2-sulfonic Acid
9AQ26S	9,10-Anthraquinone-2,6-disulfonic Acid
9AQ27S	9,10-Anthraquinone-2,7-disulfonic Acid
A3SA	Aniline-3-sulfonic Acid
A4SA	Aniline-4-sulfonic Acid
Aspa	Aspartic Acid
5Avl	5-Aminovaleric Acid
Azrd	Aziridine
Azt	Azetidine
(b)	Beta
BA	Benzoic Acid
BAzB	Brilliant Alizarine Blue
B14BQ	Monobromo-1,4-benzoquinone
25B14B	2,5-Dibromo-1,4-benzoquinone
14Bda	1,4-Butanediamine
BdGr	Binshedler's Green
Bhoa	Benahydroxamic Acid
Bla	n-Butylaniline
2B5M14	2-Bromo-5-methyl-1,4-benzoquinone
12BQ	1,2-Benzoquinone
BrAc	Bromoacetic Acid
BrBu	Alpha-Bromobutyric Acid
2BrPy	2-Bromopyridine
BsiA	Benzenesulfinic Acid

Abbreviation	Name
Bu	Butyric Acid
ByoA	(trans)-But-2-yn-1-oic Acid
Bzac	Benzoylacetone
Bzd	Benzamide
Bzde	Benzidine
Bzdl	Benzimidazole
Bzma	Benzylmethylamine
Bzm	Benzamidine
Bzs	Benzylmercaptan
Bztrz	Benztriazole
23C14B	2,3-Dichloro-1,4-benzoquinone
25C14B	2,5-Dichloro-1,4-benzoquinone
26C14B	2,6-Dichloro-1,4-benzoquinone
CbAc	Carbamoylacetic Acid
Cbma	Carbamylmethyl amine
C14BQ	Monochloro-1,4-benzoquinone
cCnA	(cis) Cinnamic Acid
Cha	Cyclohexylamine
CHC	Cyclohexane-carboxylic Acid
12CHDA	(cis or trans)-1,2-Cyclohexanediamine
Chl	Chloral
Chxm	Cyclohexanonimine
Chxo	Cyclohexanone
Cin	8-Hydroxycinnoline
Cit	Citric Acid
ClAc	Chloroacetic Acid
2C5M14	2-Chloro-5-methyl-1,4-benzoquinone
Cna	Cyanamide
CNAc	Cyanoacetic Acid
CNAct	Cyanoacetamide
CNAd	Cyanic Acid
CNea	Cyanoethylamine

Abbreviation	Name
CNMA	Cyanomethylamine
3CNPy	3-Cyanopyridine
Coy	Kojic Acid
CPC	Cyclopropane-carboxylic Acid
CPICP	m-Chlorophenol-indo-2,6-dichlorophenol
26CPIP	2,6-Dichlorophenol-indophenol
2CPy	2-Chloropyridine
3CPy	3-Chloropyridine
CsBl	Cresyl Blue
CSS	Ciba Scarlet Sulfonate
Ctch	Catechol
CX14BQ	Monochloroxylo-1,4-benzoquinone
Cy	$(Cyanide)_6$
(D)	Delta
DAcAc	Diacetylacetone
12DAP	1,2-Diaminopyridine
13DAP	1,3-Diaminopropane
Data	1,2-Diaminocyclohexanetetraacetic Acid
DC14BQ	Dicyano-1,4-benzoquinone
DClAc	Dichloroacetic Acid
DCna	Dicyanamide
DCPIC	2,6-Dichlorophenol-indo-o-cresol
Dcsa	Docosylamine
Dda	Dodecylamine
Dea	Diethylamine
Dee	Diethylether
Deta	Diethylenetriamine
Dgd	Diguanide
Dge	N,N-Dihydroxyglycine
26Dma	2,6-Dimethylaniline
Dma	Dimethylamine
DMAc	Dimethylacetic Acid

Abbreviation	Name
25DMBQ	2,5-Dimethyl-1,4-benzoquinone
DmHas	8-Hydroxy-2,4-dimethylquinazoline
DMMal	Dimethylmalorate
DM14NQ	Dimethyl-1,4-Naphthoquinone
DNP	1,2-Diaminopropane
13D2P	1,3-Diamino-2-propanol
DPcA	Dipicrylamine
DPhAm	Diphenylamine
DPK	Diphenyl ketimine
Dyp	Dipyridyl
Ea	Ethylamine
EAcAc	Ethylacetoacetate
25E14B	2,5-Diethoxy-1,4-benzoquinone
Ecea	Ethoxycarbonylethylamine
12Eda	1,2-Ethanediamine
Edta	Ethylenediaminetetraacetic Acid
EMal	Ethyl malonate
En	Ethylenediamine
Eoa	Ethanolamine
Ers	Eriochrome Blue Black R
Esa	Eriochrome Black A
Esb	Eriochrome Blue Black B
Est	Eriochrome Black T
ExCAc	Ethoxycarbonylacetic Acid
FAc	Fluoroacetic Acid
F2C	Furan-2-carboxylic Acid
F3C	Furan-3-carboxylic Acid
2FPy	2-Fluoropyridine
FumA	Fumaric Acid
Gcd	Glutaconic dialdehyde
Gl	Glycine
Glac	Glycerine

Abbreviation	Name
Glcn	Gallocyanine
Glgl	Glycylglycine
GlOH	Glycol
Glpn	Gallophenine
Glt	Glutarimide
Glta	Glutaric Acid
Glu	Gluconate
Gly	Glycollate
(gm)	Gamma
GME	Glycol methyl ether
Gu	Guanidine
Has	8-Hydroxyquinazoline
25H14B	2,5-Dihydroxy-1,4-benzoquinone
Hda	Hexadecylamine
Hed	N-Hydroxyethylethylenediamine, triacetic Acid
Hem	Beta-Hydroxyethylmercaptan
HexA	Hexanoic Acid
Himda	Beta-Hydroxyethyliminodiacetic Acid
Hox	5-Hydroxyquinazoline
HRsl	Dihydroresorcinol
(hyd)	hydrous
IAc	Iodoacetic Acid
iBu	Isobutyric Acid
IDS	Indigodisulfonate
Im	Imidazole
Imda	Iminodiacetic Acid
Imdp	Iminodipropionic Acid
Impa	Iminopropionicacetic Acid
InScr	Induline Scarlet
iPa	Isopropylamine
iPhA	iso-Phthalic Acid
2IPy	2-Iodopyridine

Abbreviation	Name
Ist(a)	Alpha-Isopropyltropolone
Ist(b)	Beta-Isopropyltropolone
ITS	Indigotrisulfonate
Lac	Lactic Acid
Ma	Methylamine
mAaBA	m-Acetamidobenzoic Acid
mAcBA	m-Acetylbenzoic Acid
Mal	Malonic Acid
maP	m-Acetylphenol
mAxBA	m-Acetoxybenzoic Acid
2M14BQ	2-Methyl-1,4-benzoquinone
23M14B	2,3-Dimethyl-1,4-benzoquinone
25M14B	2,5-Dimethoxy-1,4-benzoquinone
26M14B	2,6-Dimethoxy-1,4-benzoquinone
mBa	m-Bromoaniline
mBBA	m-Bromobenzoic Acid
mBPIP	m-Bromophenol-indophenol
mBP	m-Bromophenol
235MBQ	2,3,5-Trimethyl-1,4-benzoquinone
mC	m-Cresol
mCa	m-Chloroaniline
mCBA	m-Chlorobenzoic Acid
MCBl	Methyl Capri Blue
Mcin	8-Hydroxy-4-methylquinoline
Mcma	Methoxycarbonylmethylamine
mCNa	m-Cyanoaniline
mCNBA	m-Cyanobenzoic Acid
mCNP	m-Cyanophenol
mCP	m-Chlorophenol
2Mea	2-Mercaptoethylamine
MeBl	Methylene Blue
MelA	Mellitic Acid

Abbreviation	Name
MeOxin	8-Hydroxy-6-methylquinoline
meOxin	8-Hydroxy-7-methylquinoline
MePhs	Methyl phosphate
Mesa	2-Mercaptoethane sulfonic Acid
Met(a)	Alpha-Methyltropolone
Met(b)	Beta-Methyltropolone
mExa	m-Ethoxyaniline
mFa	m-Fluoroaniline
mFBA	m-Fluorobenzoic Acid
mFmP	m-Formylphenol
mFP	m-Fluorophenol
2M5i14	2-Methyl-5-isopropyl-1,4-benzoquinone
mIa	m-Iodoaniline
mIBA	m-Iodobenzoic Acid
mIP	m-Iodophenol
MlcA	Maleic Acid
Mln	Malonitrile
mMa	m-Methylaniline
mMBA	m-Methylbenzoic Acid
mMP	m-Methoxyphenol
mMSa	m-Methylthioaniline
mMSfa	m-Methylsulfonylaniline
mMSP	m-Methylthiophenol
mMsP	m-Methylsulfonylphenol
mMxA	m-Methoxyaniline
mMxBa	m-Methoxybenzoic Acid
mMxCa	m-Methoxycarbonylaniline
mNOa	m-Nitroaniline
mNOBA	m-Nitrobenzoic Acid
mNOP	m-Nitrophenol
mOHBA	m-Hydroxybenzoic Acid
mOxin	8-Hydroxy-5-methylquinoline

Abbreviation	Name
26M4P	2,6-Dimethyl-4-pyrone
2MPa	2-Mercaptopropionic Acid
mPhA	m-Phenylaniline
MpHas	8-Hydroxy-4-methyl-2-phenylquinazoline
mPhDA	m-Phenylenediamine
Mphl	Morpholine
mPhP	m-Phenylphenol
mPxBA	m-Phenoxybenzoic Acid
2MPy	2-Methylpyridine
3MPy	3-Methylpyridine
4MPy	4-Methylpyridine
MSAc	Methylthioacetic Acid
mSfBA	m-Sulfonbenzoic Acid
mSfP	m-Sulfonphenol
MSGly	Methylthioglycollic Acid
mSmBA	m-Sulfamylbenzoic Acid
MSOAc	Methylsulfonylacetic Acid
mTFMa	m-Trifluoromethylaniline
2MTHP	2-Methyl-$(\text{Delta})^2$-tetrahydropyridine
MtOH	Mannitol
MxAc	Methoxyacetic Acid
Mxea	Methoxyethylamine
2Mxp	2-Methoxypyridine
3Mxp	3-Methoxypyridine
4Mxp	4-Methoxypyridine
Nac	Nitroacetic Acid
NAcGu	N-Acetylguanidine
12N46D	1,2-Naphthoquinone-4,6-disulfonate
Ndap	Nitrilodiaceticpropionic Acid
NDot	N-Diethyl-o-toluidine
Ndpa	Nitrilodipropionicacetic Acid
Nea	N-Ethylaniline

Abbreviation	Name
Neal	N-Diethylaniline
Nh	$(Ammonia)_6$
NGu	N-Methylguanidine
NiPAl	N-Isopropylaniline
Nmal	N-Dimethylaniline
NMPip	N-Methylpiperidine
NMPld	N-Methylpyrrolidine
NMot	N-Dimethyl-o-toluidine
NMtAl	N-Methylaniline
NNGu	N,N-Dimethylguanidine
NN'Gu	N,N'-Dimethylguanidine
NOAc	Nitroacetic Acid
nOcA	n-Octanoic Acid
NOET	Nitroethane
NOM	Nitromethane
1Np	1-Naphthol
2Np	2-Naphthol
1NpAm	1-Naphthylamine
2NpAm	2-Naphthylamine
nPa	n-Propylamine
1NptA	1-Naphthoic Acid
2NptA	2-Naphthoic Acid
12NQ	1,2-Naphthoquinone
14NQ	1,4-Naphthoquinone
14NQ2S	1,4-Naphthoquinone-2-sulfonate
14NQ3S	1,4-Naphthoquinone-3-sulfonate
12N4S	1,2-Naphthoquinone-4-sulfonate
NSICP	1-Naphthol-2-sulfonate-indo-2,6-dichlorophenol
NSIP	1-Naphthol-2-sulfonate-indophenol
Nta	Nitrilotriacetic Acid
Ntba	N-tert-Butylaniline
Ntp	Nitrilotripropionic Acid

Abbreviation	Name
Ntrm	Nitramide
NuRd	Neutral Red
nVlA	n-Valeric Acid
Oa	Octylamine
oAaBA	o-Acetamidobenzoic Acid
oAcBA	o-Acetylbenzoic Acid
oAxBA	o-Acetoxybenzoic Acid
oBa	o-Bromoaniline
oBBa	o-Bromobenzoic Acid
oBP	o-Bromophenol
oC	o-Cresol
oCa	o-Chloroaniline
oCBA	o-Chlorobenzoic Acid
oCP	o-Chlorophenol
oCPIP	o-Chlorophenol-indophenol
18oda	1,8-Octanediamine
oExA	o-Ethoxyaniline
oFa	o-Fluoroaniline
oFBA	o-Fluorobenzoic Acid
oFmP	o-Formylphenol
oFP	o-Fluorophenol
OHAc	Hydroxyacetic Acid
OH14BQ	Hydroxy-1,4-benzoquinone
2OHQ	2-Hydroxyquinoline
3OHQ	3-Hydroxyquinoline
4OHQ	4-Hydroxyquinoline
5OHQ	5-Hydroxyquinoline
6OHQ	6-Hydroxyquinoline
7OHQ	7-Hydroxyquinoline
OHClBQ	2,5-Dihydroxy-3,6-dichloro-1,4-benzoquinone
oIa	o-Iodoaniline
oIBA	o-Iodobenzoic Acid

Abbreviation	Name
oIP	o-Iodophenol
oMa	o-Methylaniline
oMBA	o-Methylbenzoic Acid
OMiU	O-Methyl-iso-urea
oMP	o-Methoxyphenol
oMxA	o-Methoxyaniline
oMxBA	o-Methoxybenzoic Acid
oMxCa	o-Methoxycarbonylaniline
oNOa	o-Nitroaniline
oNOBA	o-Nitrobenzoic Acid
oNOP	o-Nitrophenol
oPhA	o-Phthalic Acid
oPha	o-Phenylaniline
oPhBA	o-Phenylbenzoic Acid
oPhDA	o-Phenylenediamine
oPhP	o-Phenylphenol
oPxBA	o-Phenoxybenzoic Acid
Ox	Oxalic Acid
OxAc	Oxaloacetic Acid
Oxac	Oxalacetate
Oxd	8-Hydroxy-2-methylquinoline
Oxin	8-Hydroxyquinoline
P4a	Pyridine-4-aldehyde
pAaBA	p-Acetamidobenzoic Acid
pAcBA	p-Acetylbenzoic Acid
paP	p-Acetylphenol
pAxBA	p-Acetoxybenzoic Acid
pBa	p-Bromoaniline
pBBA	p-Bromobenzoic Acid
pBP	p-Bromophenol
pC	p-Cresol
pCa	p-Chloroaniline

Abbreviation	Name
pCBA	p-Chlorobenzoic Acid
pCNa	p-Cyanoaniline
pCNBA	p-Cyanobenzoic Acid
pCNP	p-Cyanophenol
pCP	p-Chlorophenol
13Pda	1,3-Propanediamine
Pea	Phenylethylamine
Peba	Beta-Phenylethylboric Acid
2PeoA	(trans) Penta-2-en-1-oic Acid
3PeoA	(trans) Penta-3-en-1-oic Acid
pExA	p-Ethoxyaniline
pFA	p-Fluoroaniline
pFBA	p-Fluorobenzoic Acid
pFmP	p-Formylphenol
pFP	p-Fluorophenol
PgOH	Propargyl alcohol
Ph	Phenanthroline
PhAc	Phenylacetic Acid
PhBA	Phenylboric Acid
Phnz	Phenazine
PhPhA	Phenylphosphonic Acid
PhSH	Phenylmercaptan
PhT	Phthalic Acid
Phz	Phthalazine
pIa	p-Iodoaniline
pIBA	p-Iodobenzoic Acid
pIP	p-Iodophenol
Pip	Piperidine
Pipz	Piperazine
Pld	Pyrrolidine
pMa	p-Methylaniline
pMBA	p-Methylbenzoic Acid

Abbreviation	Name
pMCP	p-Methoxycarbonylphenol
pMP	p-Methoxyphenol
pMSfa	p-Methylsulfonylaniline
pMSP	p-Methylthiophenol
pMsP	p-Methylsulfonylphenol
pMTa	p-Methylthioaniline
pMxA	p-Methoxyaniline
pMxBA	p-Methoxybenzoic Acid
pMxCa	p-Methoxycarbonylaniline
Pn	Propylenediamine
pNOa	p-Nitroaniline
pNOBA	p-Nitrobenzoic Acid
pNOP	p-Nitrophenol
pOHBA	p-Hydroxybenzoic Acid
pPhA	p-Phenylaniline
pPhDA	p-Phenylenediamine
pPhP	p-Phenylphenol
PplA	(trans) Propiolic Acid
pPxBA	p-Phenoxybenzoic Acid
Pr	Propionic Acid
Pr2C	Pyrrole-2-carboxylic Acid
pSfBA	p-Sulfonbenzoic Acid
Psfn	Phenosafranine
pSfP	p-Sulfonphenol
PSIBP	Phenol-m-sulfonate-indo-2,6-dibromophenol
pSmBA	p-Sulfamylbenzoic Acid
pTFMa	p-Trifluoromethylaniline
Ptn	1,2,3-Triaminopropane
pTSiA	p-Toluenesulfinic Acid
Py2C	Pyridine-2-carboxylic Acid
Py3C	Pyridine-3-carboxylic Acid
Py4C	Pyridine-4-carboxylic Acid

Abbreviation	Name
Pydz	Pyridazine
Pymd	Pyrimidine
PyNO	Pyridine-N-oxide
Pyr	Pyridine
Pyrr	Pyrrole
Pyzl	Pyrazole
Pyzn	Pyrazine
Qn	Quinoline
Qnxl	Quinoxaline
Rsl	Resorcinol
RsSf6	Rosindone Sulfonate No. 6
Saa	2-Sulfoaniline, diacetic Acid
SafT	Safranine T
Sal	Salicylic Acid
Sald	Salicylaldehyde
Sccm	Succinimide
SCNAc	Thiocyanatoacetic Acid
sDPGu	sym-Diphenylguanidine
SMiT	S-Methyl-iso-thiourea
SP2C	Thiophene-2-carboxylic Acid
SP3C	Thiophene-3-carboxylic Acid
2SQ	2-Mercaptoquinoline
3SQ	3-Mercaptoquinoline
4SQ	4-Mercaptoquinoline
SSald	Sulfosalicylaldehyde
Suc	Succinic Acid
TAcM	Triacetylmethane
Tart	Tartaric Acid
Tate	Triaminotriethylamine
tBa	tert-Butylaniline
2tBa	2-tert-Butylaniline
tBPy	2-tert-Butylpyridine

Abbreviation	Name
TB12BQ	Tetrabromo-1,2-benzoquinone
TC12BQ	Tetrachloro-1,2-benzoquinone
TClAc	Trichloroacetic Acid
tCnA	(trans) Cinnamic Acid
Tea	Triethylamine
Teta	Triethylenetetramine
Tf	Theonyltrifluoroacetone
TFAc	Trifluoroacetic Acid
TGa	Thioglycollic Acid
Thzl	Thiazole
TlBl	Toluylene Blue
Tma	Trimethylamine
TMAc	Trimethylacetic Acid
Tmen	Trimethylenediamine
TMPDA	N,N'-Tetramethyl-p-phenylenediamine
TMPsfm	Tetramethylphenosafranine
Tmta	Trimethylenediamine, tetra acetic Acid
TNOBA	2,4,6-Trinitrobenzoic Acid
TNPh	Trinitrophenol
TPhA	Terephthalic Acid
Trna	Triethanolamine
Trop	Tropolone
TUr	Thiourea
Ur	Urea
Uda	Undecylamine